Fisheries Research Planning and Management in Developing Countries

Dr. V. R. P. Sinha
Ph.D. (Liverpool), FNA, FNASc, FNAAS

Founder Vice-Chancellor
Central Institute of Fisheries Education,
Mumbai, India
and
Former Consultant,
Food & Agriculture Organization of the United Nations
and the World Bank

Narendra Publishing House
DELHI

ISBN: 81-85375-92-5

First Published 2005

Published By:

Narendra Publishing House

PUBLISHERS AND DISTRIBUTORS

1417, Kishan Dutt Street, Mali-Wara, DELHI- 110006 (India)

Ph. : 91-011-23268470 ● Telefax : 91-011-23259412

Email : info@nph-india.com *www.nph-india.com*

Acknowledgements

Gratitude for help received in writing the book is acknowledged to all those who contributed in any way. Special mention is made of the Indian Council of Agricultural Research New Delhi, India for providing me the opportunity establishing the Central Institute of Freshwater Aquaculture. Bhubaneswar and other associated institutions and for up gradation of the Central Institute of Fisheries Education to the Deemed University, Mumbai India. Thanks are also due to the Food and Agriculture Organizations of the United Nations, the World Bank, Winrock Agriculture Research Institute, Arcadis Euro Consult and the People Republic of Bangladesh for offering me various assignments dealing with fisheries research planning and management. These assignments gave me a lot of insight into different aspects of fisheries research planning and management as I could see in many developing countries.

Thanks are due to Dr. M. A. Mazid, Director General of the Bangladesh Fisheries Research Institute Bangladesh, who devoted considerable time in discussing various practical difficulties in managing the research. Thanks are due to Professor Akhouri, A. N. Sinha, University of Minnesota and VA Medical Center, Minneapolis, USA, Dr. C. Prasad, Retired Deputy Director General of ICAR, India, Mr Pratap Narain, Retired Senior Statistician FAO, Rome, Italy, Dr G.S.P.Verma Retired Professor of Chemistry, Ranchi University, India and Mr Charles Quinn, Northern Tier Regional Library, Gibsonia, USA for going through the manuscript and for their valuable suggestions.

Special mention is made of the help rendered by Dr S. Ayyappan, DDG, Fisheries, ICAR, New Delhi, India and Dr. S. Mukherjee, Director CIFE, Mumbai and his staff particularly Dr. K. Pani Prasad, Dr. W.S. Lakra, Dr D.N.Singh, Dr.R.S.Biradar Dr. K.K. Jain,

Dr. M.P. Singh Kohli, and Dr. Purushothaman. Thanks are also due to Dr.Amit Prasad of Gibsonia,USA and Mr. Vidyut Sinha of Wexford, USA for the facilities to complete the book. Dr. M.G.Hussain, Director Research, BFRI, Bangladesh Dr. D. Nath and Dr U.Bhowmick of CICFRI and Dr. S.N. Gajbhiye, National Institute of Oceanography Goa, India, deserve great appreciation for their offer of relevant photographs. Last but not the least, thanks are also due to my wife Mrs Asha Sinha for her constant encouragement.

Content

Preface

Fisheries science is of recent origin compared to other agricultural and animal sciences concerned with food production. It deals with very complex aquatic ecosystems and diverse lives of fish, crustaceans, mollusks and seaweed of different habit and habitat. The techniques of their biological production, harvesting, and post harvesting have no parallel with any other farming system in the world. Though, fisheries promise to contribute a vital share to the global food basket, the present fish demand and supply gap is widening because of human and animal population growth, domestic, agriculture and industrial pollution, competing demands for the water, man made changes in the waterways and many other constraints. All of these may reduce potential growth in fisheries. The future is also uncertain due to unknown consequences of weather aberration and global climatic change on fish habit and habitat. To obviate or mitigate these adverse effects, pragmatic fisheries research is needed.

The research involves planning, designing and managing experimentation on physico-chemical and biological status of fisheries resources, their manipulations and control and other types of inquiries involving environmental, economic, legal and social sciences, all make a continuum of scientific studies from basic research through applied and adaptive research. It is important that research problems are correctly identified, prioritized and studied to achieve meaningful results to achieve the optimal potentiality of fisheries, consistent with national priorities, fish farmers and fishermen needs, prevailing environmental and economic conditions in the country and institutional capacities.

The planning process becomes a very crucial and vital activity in fisheries research requiring large physical and financial resources, which are seriously limited world over. Also, uncertainty exists about the availability and assured continuity of these resources. Implementing the plan and managing research involve both scientific and financial performance, audit and accounting, management and motivation of scientists and support personnel, providing leadership and training of personnel and arranging support services. And then, disseminating the results of research by proper documentation and publication along with extension and demonstration, education and training, all of which are the integral parts of research management.

With the above background the book attempts to outline the nature of fisheries research, its administration and organization of research institutions along with delineation of fisheries and aquaculture disciplines. It also highlights the major issues confronting sustainable fisheries development in developing countries, and provides a framework of research planning and management along with packaging and transfer of technology, capacity enhancement in fisheries education, extension and training.

— V R P Sinha

1. Introduction

1.1 Nature of Fisheries Research

Fisheries deal with aquatic organisms management and procurement from open water and with farming of such organisms under controlled conditions. Though the art of fishing is known since the dawn of civilization, fisheries science as such is of recent origin compared to other agricultural and animal sciences concerned with food production. Nonetheless, it is becoming a dominant and important science with a promise to contribute a vital share to the global food basket. It deals with very complex, multi-dimensional aquatic ecosystems of oceans, seas, estuaries, rivers, reservoirs and ponds and with diverse lives of fish, crustaceans, mollusks and seaweed of different habit and habitat. The techniques of their biological production, harvesting, and post harvesting have no parallel with any other farming system in the world.

Fisheries fall under two main categories: capture and culture (commonly known as aquaculture). Capture fisheries in the open water are very complex involving different fish population, their interaction, feeding, breeding, and migration in relation to the existing variables relating to oceans, seas and rivers ecosystem. They are exploited as a common property resource. Aquaculture, on the other hand is farming of aquatic organisms under controlled conditions (to a certain extent) and with ownership right. Its productivity is governed in general by the principle of 'as you sow so you reap' and its management is highly labor and input dependent.

Primary productivity, whether on the land or in the water, is a process of energy transformation. In the aquatic ecosystem, the solar energy is fixed by chlorophyll bearing plants predominantly algae and aquatic weeds; the primary producers, which are consumed by aquatic herbivores. Usually, different forms of life are linked together in the food production and consumption through a general pattern of food web i.e., primary producers-herbivores-carnivores. However, this straight food chain appears simple, but in fact, it is more complex in an aquatic medium than in a terrestrial ecosystem.

The aquatic ecosystems are highly diverse and so are their characteristics. Even water in a pond appears as static, but it is highly dynamic - changing its chemistry day and night. A seasonal pond (i.e. one whose bottom is exposed to sunlight for a few months in a year) has different physico-chemical characteristics than a perennial pond. The running water of a river is different than the water in an impoundment or in an estuary. Its quality gets affected by changes in the rivers' catchments and by any other human intervention, whether it is in the up hill or in the river or in the farms and factories or down in the ocean depth. The marine ecosystem with tides and currents is nonetheless highly complex.

The changes in the aquatic environment and also in the morphology of aquatic organisms are as such less visible than those in the terrestrial ecosystem and the crops and animals it supports. These create difficulties for the researchers in their studies on the one hand and development workers on the other to raise public awareness of the environmental issues in an aquatic environment.

The ever-increasing demands of water that are pressing against hydrological limits have an alarming effect on the aquatic resources and on the ecosystems. Conversion, diversion and fragmentation of wetland and other water bodies are seen all over the world. Freshwater consumption by human population increased six fold during last century and over two dozen countries are declared water stressed and many more will be so in the future (Postal 1991). Water management for irrigation, flood control and industrial developments in many countries have directly affected the quality of aquatic resources and their biodiversity. In fact, of all the threatened lives on earth, aquatic organisms have been the most affected.

The situation is further aggravated because of aquatic pollution, caused by agrochemical and other industrial pollutants. Man-made climatic change through global warming further threatens the habit and habitat of the aquatic organisms. The erratic and extreme climatic events and particularly intense monsoon with damaging floods are predicted which add to the threat.

The fisheries resources are renewable but they are being subjected to over exploitations. Swaminathan (1996) projected that by 2010 the availability of renewable resources such as fish catch, irrigated land, cropland and forest compared to population size will show negative per capita availability. The fishing fleets are 40% in surplus capacity vis-à-vis what the seas and oceans can sustain (Linden 2000). Fisheries of many important species have already collapsed and many more will follow suit, unless management decisions are made and implemented on the basis of the results of research on the dynamic changes taking places involving biological, environmental, economic, legal and social aspects of fisheries.

It is expected that the present fish demand and supply gaps will further increase due to human and animal population growth, pollution and siltation of the water bodies, competing demands for the water from different sectors of economy and the over-riding but unknown consequence of weather aberration and global climatic change on fish habit and habitat. To obviate or mitigate these adverse effects, pragmatic fisheries research is needed , that is of course difficult and complex but is exciting. It is full of opportunities and constraints. The complex issues that are expected to be answered by the fisheries scientists are many and thus their task is difficult and enormous.

1.2 Fisheries Sector and its Role in National Economy

Fish is the primary source of protein for over 1 billion people in the world and fishing is the main source of income of over 100 millions who are directly involved in fishing. Fish is very nutritious and more healthful than other meat.

Apparently fisheries sector is small in terms of volume but it significantly contributes to national income, nutritional security, export earning and in fulfilling social objectives, with a great promise to contribute further.

The trend in global fish production shows that over the last three decades marine sector in developed countries show very poor performance and even negative growth. The production has gone down in the last decade from 32 million tons to 30 million tons, showing a negative growth @ 2.5%. Similarly, the inland sector is stagnating at 1.5 million tons showing a negative growth @ 3.5% during the same period. However, bright pictures have now emerged from the developing countries. During the last three decades marine sector increased its contribution from 27.3 to 70 million tons and inland from 4.6 to 28.7 million tons, showing a very impressive growth @ 4.5% and 8.7% respectively , much of this has been due to increased aquaculture production (table 1).

Since 1970 aquaculture production worldwide has increased at an average compounded rate of 9.2 % per year compared to 2.8 % for meat production. It has been one of the fastest growing food production sectors over the past two decades in many countries. In fact, it grew more than five times as fast as the global population and it is expected to grow with similar rate in near future.

Table 1 - Global Fish Production Based on FAO (2002)

World/ Production Region (million MT)				Average annual rate of growth (%)			
	1970	**1980**	**1990**	**2000**	**1970-80**	**1980-90**	**1990 2000**
World							
Marine	59.4	64.7	84.1	100.2	0.9	2.7	1.8
Inland	6.0	7.7	14.5	30.2	2.5	6.5	7.6
Total	**65.4**	**72.4**	**98.6**	**130.4**	**1.0**	**3.1**	**2.8**
Developed countries							
Marine	32.1	37.2	38.9	30.2	1.5	0.4	-2.5
Inland	1.5	1.7	2.1	1.5	1.3	2.2	-3.0
Total	**33.5**	**38.9**	**40.9**	**31.7**	**1.5**	**0.5**	**-2.5**
Developing countries							
Marine	27.3	27.5	45.2	70.0	0.1	5.1	4.5
Inland	4.6	6.1	12.5	28.7	2.8	7.5	8.7
Total	**31.9**	**33.6**	**57.7**	**98.7**	**0.5**	**5.6**	**5.5**

Freshwater aquaculture has a long tradition in many countries of Asia and is an important component of fisheries. Asia contributes about 84% of the global production as compared to 8.5 % by Europe, 3.7 % by North America, 2.3 % by South America and about 0.5 % by Africa .

Fish culture in Latin America is of fairly recent origin: the most commonly farmed species are exotic to the continent. Culture of carp, salmon and trout is insignificant and tilapia is restricted to few countries. Aquaculture in Africa also has unfortunately, been lagging behind. Except in a few countries like Cote d' Ivoire, Egypt, Liberia, Madagascar, Nigeria, South Africa, Tanzania, Zanzibar, Zaire, Zambia and Zimbabwe, aquaculture is still in infancy.

Since 1980s the coastal aquaculture has attracted the global attention in view of the surging shrimp demand in developed countries. The traditional practice of 'trapping and holding' operation of shrimp and fish seed in coastal areas of many Asian countries provided the basis of intensification of shrimp culture in those countries. They made heavy investment in developing infrastructure and took initial advantage of the global trade opportunity. But the success could not last long because of over enthusiasm of entrepreneurs in indiscriminate expansion and intensification of shrimp farming. Shrimp disease epidemic occurred and eventually shrimp farming ran into trouble. However, it is being revived after learning the painful lesson of loss and devastation.

Aquaculture in coastal and freshwater has distinct roles to play in national development. Supplying cheap animal protein food to the poorest and undernourished segment of the rural population should be one of the highest priorities of freshwater aquaculture in inland areas, whereas to earn foreign exchange and to provide additional employment should be in coastal aquaculture. Thus, rural freshwater aquaculture should focus on producing cheap food crop and coastal aquaculture on producing cash crop. Coastal aquaculture has outstanding advantage to those countries with high population density. It opens up new frontiers to displaced fishermen and does not compete with crop production for limited arable land resources.

Mariculture also popularly known as "Sea farming" is highly developed in Japan, Republic of Korea, China, the Philippines and Taiwan (China) particularly for seaweed and mollusk culture. Chile has well developed seaweed culture and processing industry. But other countries in South America have not done much. So, is in the case with African countries, where mariculture is only prevalent in Tunisia, South Africa, Morocco, Seychelles and Zanzibar.

International trade in fish and fishery products had been forging ahead till recently. But due to general downturn of economy in certain developed countries, the trade is facing some degree of turbulence worldwide. Yet, about 49.4 million tons of fish and 4.5 million tons of seaweed (of the total world fish production of 130.2 million tons and of seaweed of 11.9 million tons) were traded internationally in 2001. The total value of export was US$55.9 billion f.o.b. (free on board), whereas total import was US$ 59.4 billion c.i.f. (cost, insurance & freight). Export was 1.1% more and import was 1% less compared to 2000 (Vannuccini 2003). According to the latest FAO report (2002), the 86 Low- Income Food-Deficit Countries (LIFDCs) accounted for 19% of total exports in value terms. The developing countries continue to supply about 50% and the developed countries the rest 31%. The net receipts of foreign exchange in fishery trade by the developing countries showed an increase from US$ 4.0 billion in 1981 to US$ 17.7 billion in 2001, an increase to 400% during last two decades. The amount was greater than the net exports of other agricultural commodities such as rice, cocoa, tobacco and tea. Therefore fisheries sector has become very important in many developing countries and is viewed as an important component of the national development plans.

1.3 Fisheries Research Planning and National Policy

Aquatic resources are vast and varied and many investigations need to be of long- term duration and at times of continuous nature. They require large physical and financial resources and manpower, which are seriously limited in developing countries. Also uncertainty exists about their availability for longer duration. Under such constrained situation research planning process becomes very crucial so that urgent research problems are correctly identified, prioritized and studied to achieve meaningful results to harness the optimal potentiality of fisheries.

Planning needs to be responsive to changing social, economic, legal, environmental and other conditions, and the research plans need to be re-evaluated and adjusted in light of such changes. Research planning must be realistic and should take into account of the available human, financial and physical resources. Planning should be flexible and dynamic to take advantage of new opportunities and promising directions that emerge from the on-going work. The research plans need to be 'need-based' and also 'opportunity-based'.

It is essential that research activities are consistent with national priorities, fish farmers and fishermen needs, prevailing environmental and economic conditions in the country and institutional capacities. Therefore, every researcher should be fully familiar with the National Fisheries Policy and Plan. Equally, it is important that national water planning and environmental aspects are also kept in mind because they are highly relevant and related to fisheries development.

Basically Fisheries Policy of different developing countries revolves around rational exploitation of fisheries resources, their conservation and development and in particular improving socio-economic conditions of fishermen and fish farmers and in creating entrepreneurship. It is also intended to meet the national animal protein requirement and to earn foreign exchange through export of fish and fishery products.

The National Environment Policy and Water Policy of the developing countries should match with the Fisheries Policy particularly in environment friendly development of water resource. The policies should intend to control all kinds of aquatic pollution, arrest further degradation of aquatic environment and also to undertake eco-restoration and rehabilitation of aquatic resources.

Water is no more taken for granted and now it is realized that water is a scarce and precious resource, which needs conservation and development in an integrated manner. Many countries initiated national water planning for water resource conservation and development. It is essential to effectively manage abundant water in rainy seasons and scarce in other seasons, to avoid flooding and drought, which directly affect fisheries resources and fish lives. While economizing use of water in agriculture and industry with proper water management technologies, it is essential to ensure equitable distribution of water to aquaculture and fisheries along with different sectors of economy.

Agriculture emerged as the most consumptive sector of water. Most of the developing countries have made heaviest investment in creating irrigation facilities for increasing agricultural production without caring much of its effect on fish biodiversity or fisheries resources. But of late, the issues of environment, equity and economic competitiveness have appeared on the center stage of water resources development programmes (Swaminathan 1995). Agriculture production planning in many countries is now giving highest priority for an efficient irrigation and striking a balance with ecology and social issues with increasing productivity. Similarly, governments are showing greater concern of aquatic pollution by the industry and are trying to obviate it through promulgation of anti pollution laws.

With the full background information on country's national policy about fisheries, agriculture, industry, environment etc., the fisheries researcher and research managers should study the fisheries resource potential, issues and constraints to develop fisheries with environmental sustainability and social equity. They should evolve a strategy and a vision of research to fulfill the objectives of their study.

It is cautioned that research planning can achieve the desired objectives only when the research studies are undertaken with sincere commitment and matched with timely input delivery, meticulous monitoring and management of research investigations, packaging of research results for technology generation and management guideline formulation. And then, of course transfer of technology and effective implementation of guidelines for management and their follow up by the government development agencies. It needs to be appreciated by all concerned that research is only the basic requirement and one of the components of development or management but effective mechanisms are essential for the use of research outputs. There are many examples of good research findings that have little impact on fisheries development because of the lack of effective mechanism for the use of such results (World Bank 1991).

Managing research involves both financial and scientific performance, audit and accounting, administrating and motivation of scientists and support personnel, providing leadership and training and arranging support services. Lastly, disseminating the results of research through proper documentation and publication along with extension and demonstration, education and training are vital and the integral parts of the management.

It is important that research managers should understand and appreciate that the essence of management is making a decision and the essence of good management is making a better decisions. Management, in fact, is the art and science of making right decision at the right time for a successful completion of a particular endeavor.

2. Basic Terms in Research Planning

2.1 Terms in Research Planning

Planning and managing fisheries research involve planning, designing and managing experimentation on physico-chemical and biological status of fisheries resources, their manipulations and other types of inquiries involving environmental, economic, legal and social aspects These require scientific studies from basic research to applied and adaptive research.

2.1.1 Mandate of the Institution

All research institutions have a past history of the establishment and over the time they grow and thus they are at different levels of development. What is seen at one point of time is the stage of development of the same tree, which once upon a time was planted. But sometimes the history is forgotten and thus there is a likelihood that the very purpose of establishment of the institution gets diluted or diffused. History plays a crucial role in reminding the workers about the quantum of efforts put in by the past workers to bring the whole organization into being. Appreciation of which is important to enthuse the staff to work under any difficulty for the purpose of upholding the socio-economic commitment for which the institution was established.

It is always important to know the exact mandate of the institute with the broad statement of purpose about it, why it was created, what it does and why it does it? This statement is commonly known as Mission Statement. It defines the boundaries of the organization's territory, and indicates its commitment to the society. The mission remains more or less the same over a period, however, programs change from time to time to solve the emerging constraints and also to take advantage of the opportunities in the sector.

2.1.2 Goals of research studies

The research project that the institute undertakes, should have identified goals, which should be well defined and clear. Goals indicate the anticipated outcomes of each research project in time bound manner and their quantifiable impacts on the society (FAO/UNDP 1990).

The goals of any research studies could be economic, social, environmental, scientific, or a combination of all. For example, economic goals could be to increase aquatic productivity, or increased contribution to national food security, or increased foreign exchange earning. Similarly, social goals could be to increase participation of women in fisheries or to increase rural and coastal employment etc. Environmental goals could be the protection of endangered species, or maintenance and protection of water quality, etc. Similarly, scientific goals could be to understand the scientific basis of complexities of aquatic resources.

In turn the goal is supported by specific objectives; for examples, under the goal of "Protection of endangered species" research objectives may include:

- Development of breeding technology of 20 critically endangered species;
- Development of guidelines for declaration of a biosphere and earmarking a particular stretch of the ecosystem as a reserve for breeding and raising of larvae of threatened species
- Development of legal framework for patenting and registering genetic resources of fish and other aquatic resources.

2.2 Kinds of Research

There are different types of scientific inquiry or research ranging from basic to applied to adaptive or on-farm. Though there is no absolute or clear-cut boundary between them, it is desirable to understand the basic differences.

2.2.1 Basic research studies

The basic research tries to understand the biological basis and process to provide knowledge and insight that are necessary to undertake wide variety of applied research. For example, studies on hormonal control of fish reproduction will help better control on their breeding. Similarly, research on nutritional requirement of fish will be required to feed them artificially.

2.2.2 *Applied research studies*

Unlike basic research, that is fundamental in nature, applied research mainly solves specific problems and generates knowledge that can be a basis for evolving the package of practices. Thus, studies on controlled breeding of carp or supplementary feeding of fish with artificial diets or fertilization of pond to enhance productivity are examples of applied research. It may be defined as to solve specific problem and to generate knowledge that can be used in practice in fisheries management.

2.2.3 *Adaptive research studies*

Applied research results form the basis of adaptive research, which mainly involves testing and modifying the results or the package of practices for adoption at a specific situation or an environment. Quality seed production of carp in different agro-climatic regions of a country becomes an adaptive research, which may involve certain modifications in the basic technology of seed production according to the varying physico-chemical conditions of water, weather conditions of the region and their effect on brood stock maturation and performance. Similarly, studies undertaken on feed formulation of carp in different agro climatic regions of a country based on locally available raw materials or for formulation of pond fertilization programs in different agro climatic conditions are clear examples of adaptive research. Thus, adaptive research can be defined as to modify the package of practices to make it suitable for a particular situation or an environment for its easy adoption. Such studies are undertaken invariably with farmers' participation on their farms and are also termed as on-farm research.

2.2.4 *Farming System research studies*

Farming System research is a holistic approach to improve the productivity of farmers' resources. Integrated farming of crop, livestock and fish serves as an example of Farming System Research studies, which are undertaken to improve the productivity of the total resources of the farmers in terms of optimal utilization of land, labor, water and waste through integration of different farming systems.

2.2.5 Strategic research

Strategic research is more on policy issue, aimed at evolving research strategies and defining priorities areas in which specific applied research projects should focus. For example, research efforts should be focused either on enhancement of pond productivity or on conservation of aquatic biodiversity or on aquatic habitat restoration or all the three areas in the prevailing circumstances and priorities set by the national policy. Strategic research addresses pressing contemporary and crucial problems.

2.3 Research Program, Project and Individual Studies

Research programs are broad areas of investigation such as aquatic productivity management for optimal fish production in confined water or in open water , the former pertains to culture related fisheries, whereas the latter to open water fisheries. Under the programs many research projects are identified which may be to evolve technology for productivity enhancement, or for environmental, social and economic sustainability. Under the projects many individual research studies are undertaken on specific problems, the results of which form the basis of packaging of the practices of a technology. The following examples will clarify the difference between program, project and individual studies.

Example No. 1

Program	**Aquatic productivity management for optimal fish production in confined water.**
Research Projects	· Study the relationship between primary productivity and fish production under different management technologies. · Study the mechanism of nutrient release from pond sediment to optimize productivity. · Study to control of eutrophication and take advantage of higher primary.productivity for optimal fish production.

Individual Studies	• Productivity monitoring of an undrainable pond. • Productivity monitoring of a drainable pond. • Role of different carps in utilizing primary and secondary productivity. • Role of different carps in improving soil and water conditions. • Role of common carps in recycling of nutrient from pond bottom. • Water quality requirement of different carps. • Role of silver carps in controlling eutrophication.

Example No. 2

Program	**Open-water fisheries management**
Research Projects	• Study the relevant data on inland fish and fisheries resources. • Study on the population dynamics and stock assessment of commercially important inland fishes. • Study on effects of human intervention on estuarine fisheries.
Individual studies	• Biology and ecology of Hilsa. • Population dynamics and recruitment pattern of Hilsa. • Grey mullet distribution, migration and feeding. • Effect of UV B on propagation of Hilsa.

3. Organization of the Fisheries Research Institute

3.1 Administrative Control

3.1.1. Research Responsibility

Historically, fisheries research has been primarily concerned with fish taxonomy and fish biology and thus it was basically one of the activities of the Zoology Department of universities. However, serious attempt to increase food production, particularly during and immediately after the World War II, brought impetus to intensify research on different aspects of fisheries in many countries including Great Britain. It was only in the later part of the 20th century that an organized effort was started in many countries to establish fisheries research set up under the department of Agriculture, or under other departments such as Forestry, Animal Husbandry, Natural Resources, Industry, Science and Technology etc.

A separate Ministry or Department to exclusively look after the fishery sector does not exist in many countries. However, the significant contribution of fisheries for the last few decades influenced many national governments to establish research units and also to organize administrative and development departments for the sector. Many international agencies particularly the Food & Agriculture Organization of the United Nations, the World Bank, and governments of Denmark, Norway, Holland, U.K., Japan etc. complemented the national efforts to establish and strengthen the research organizations and institutions.

It is normally seen that the responsibility for fisheries research is, with certain exceptions, accepted as the government responsibility (Kesteven1966). Generally, the research responsibility has either in or attached to certain executive department. For examples, in the USA, the execution lies in the department responsible for fisheries, i.e., Laboratories of U.S. Fish and Wildlife Service to the Department of Interior. In Canada, research responsibility is attached to the department responsible for fisheries i.e., Fisheries Research Board of Canada under the Minister of Fisheries. In Norway, research responsibility lies with the Norwegian State Institute of Whale Research of the Ministry of Industry. In some countries responsibility lies with the agricultural research organizations, for examples, in India, Indian Council of Agricultural Research, which is attached with the Department of Agricultural Research and Education, Ministry of Agriculture. Similarly, Council of Scientific and Industrial Research Organization in Australia looks after the execution of research. However, in many developing countries there exists administrative tug of war between various government departments to be involved in administration of fisheries research and development (World Bank 1991).

The international responsibility normally is with the regional and world organizations such as Food and Agriculture Organization of the United Nations (FAO), United Nations Educational, Scientific and Cultural Organization (UNESCO), the Southeast Asian Fisheries Development Center (SEFDEC), the Network of Aquaculture Centre of Asia (NACA) etc. or through funding and supplementing the research activities of the different research institutes, such as International Center of Living Aquatic Resources Management (ICLARM) (now World Fish Center). There are various international fisheries bodies which are involved in collaboration and cooperation among the different nations for fisheries development and also indirectly for policy related to research to some extent.

Association of Southeast Asian Nations with the membership of Brunei, Darussalam, Cambodia, Indonesia, the Lao People's Democratic Republic, Malaysia, Myanmar, the Philippines, Singapore, Thailand and Vietnam, cooperate among each other on aquaculture and fisheries development, post harvest technologies and on quality assurance for fishery products (FAO 2000). Similarly, other organizations like Caribbean Community and Common Market (CARICOM) promotes the development of fisheries of member states for optimal exploitation on sustainable basis. Commonwealth of Independent States (CIS) is restructuring their fleets and also collaborating on fish processing and marketing sectors.

The Economic Community of West Africa undertakes collaboration for fisheries development in the member states. The European Community concentrates more on conservation and responsible fisheries. Its collaborative efforts are channeled through active memberships of the Northwest Atlantic Fisheries Organization (NAFO), the Commission for the Conservation of Antarctic Marine Living Resources (CCAMLR), Northeast Atlantic Fisheries Commission (NEAFC), the Indian Ocean Tuna Commission (IOTC), the North Atlantic Salmon Conservation Organization (NASCO), the Fisheries Committee for the Eastern Central Atlantic (CECAF), the International Baltic Sea Fishery Commission (IBSFC), the General Fisheries Commission for the Mediterranean (GFCM) and the International Commission for Conservation of Atlantic Tunas (ICCAT). The Southern African Development Community (SADC) member states collaborate on marine and inland fisheries sustainable development. The South Pacific Forum (SPF) through its South Pacific Forum of Fisheries Agency (FFA) concentrates on conservation and optimal utilization of fisheries resources and oversees coastal states' sovereign rights over the living marine resources (FAO 2000).

Non-governmental organization, like universities and colleges devoted to fisheries, also execute research, and so do the non-university-learned societies, associations and industries in different parts of the world. At times fisheries research is undertaken by private sector in certain developed countries but it has not yet substituted to any great extent the government sponsored programs. However, government continued support for research is absolutely essential in developing countries, because the results of the research are public goods and private agencies do not have full incentive to invest in such goods because they cannot receive full and unshared return to their investment.

Similarly, execution of research in agriculture is done either by the government departments or by autonomous bodies or by private sector in different countries. In many developing countries the responsibility of research of the Ministry of Agriculture is shared by many other Ministries. A review on the characteristics of agriculture research systems in 12 countries in the developing world showed that only half have retained their present system of linkage of the research organization with a particular Ministry more than four years, others changed frequently, that adversely affected the direction and administration of research and also the morale of the staff (FAO/UNDP 1984).

Institutional control or structure is not in itself a determinant of the degree of effectiveness or efficiency but it is the support of government and administrators, which really matters. For research workers an important consideration is that how far the rules of decision making are vested in institute and with them. There has to be autonomy to avoid excessive hierarchical bureaucracy but certainly regulatory rules and procedures of the nodal ministry or administrative department cannot be disposed off for proper accountability.

Management of fisheries research whether carried out in fishery institutions, or in universities or in independent laboratories, should aim at to steer, promote and streamline research activity so as to avoid duplication and diffusion of efforts to maximize its results for the benefit of the sector. This should be the guiding principle of functioning of any fisheries institution.

It is important to see that extension set up, which puts the research and technology into effect is not widely separated from research organization. Appreciation of the potential contribution of research to the sectoral development is most essential and at the same time extension and development work should get the due credit.

3.1.2 Autonomy of the Institution

Institution as an autonomous body should be the most effective arrangement. However, autonomy does not mean that an institution can do whatever it wants with whatever manner or way it chooses. In fact, it places more responsibility on the institution to evolve in-built flexibility in the rules and regulations to suite the needs and urgency of research. It is highly desirable that research organization should be made autonomous with well- defined responsibility and accountability, without any diffusion of responsibilities or confused line of communication and control.

Obvious advantages exist in favor of delegating operational responsibility for research to autonomous organizations. The system should involve all interested parties and particularly stakeholders through a governing body including representatives from research, planning, extension, fishing, fisheries industry, NGO, and politics. The governing body should set guidelines for administration and management of the research organization and also ensure necessary resources for the institution. In certain developing countries political compulsion and interference in recruitment, promotion and placement of staff, are alarmingly increasing which can also be circumvented if institution functions as a fully autonomous body.

Autonomy ideally requires that the government should delegate full functional authority to the Governing Body of the institution. This in turn should delegate the power to the Chief Executive, who should give required power to the heads of the department and such delegation of power is given to all scientists down the line according to the requirement of research. However, autonomy places greater demand of accountability on the scientist and scientific managers for efficient conduct of research and disciplined financial management. They need to be fully conversant with proper administrative and financial rules, regulations and control. In fact, it is seen sometimes that enough delegation of power has been given to the scientific managers but they do not avail such powers either because they are not conversant of their power or they are not equipped with the knowledge of rules and regulations as to how to use them.

Sometimes, it is seen that certain autonomous research institutions have failed to deliver. This was perhaps because the mechanisms to achieve the objectives of the institution as set out in the act of establishment are lacking effectiveness or the required power has not been delegated to the institution or to the Governing Body. Sometimes, the Governing Body also lacks the appreciation and intricacies of research and research management. It is commonly seen that such organizations strictly follow the personnel policies like a government department with over regulation of financial management. Their rules and regulations have not been modified to provide a scientific environment of facilitating fisheries research , which is highly season sensitive, and needs timely research support in terms of facilities and inputs supply.

3.1.3 Governing Body

The constitution of the governing body depends on the existing bureaucratic set up of individual country. However, it should consist of the representative of all the stakeholders, along with chief financial authority, chief scientific administrator, chief of development and extension, distinguished university professor and headed by the chief of national development planning. Its duties should encompass the following:

- Provide policy guide-lines for undertaking research relevant to the national development plans and strictly overview that research programs reflect national objectives and priorities.
- Provide policy guide-lines for improving the quality of research.
- Make aware the government of potentiality of research in terms of national development and obtain adequate funding support.
- Provide policy guide-lines for efficient management and administration of the institute and overview their implementation.
- Approve cost – effective budget for the Institute.

- Evolve suitable personnel policy to attract the scientists and other staff.
- Approve staffing and infrastructure development.
- Approve international cooperation keeping in mind whether such cooperation serves national development objectives and priorities.
- Avoid duplication of research and research related activities in the Institute.
- Delegate appropriate authority to the Chief Executive of the Institution to function freely without fear and favor with full accountability for the performance of the Institute and the execution of research.
- Make aware the government of relevant research results and overview the use of such results.

3.2 Organigram of the Institute

It is absolutely necessary for the institute to firm up its organizational setup. A firm organizational set up helps institute administration and the scientists to perform well without overlapping or duplicating efforts of other scientists. Dilution and diffusion of research efforts are commonly seen in the institutions, which are loosely organized without giving in depth consideration of the functions of the divisions.

This results in unnecessary research expenditure and loss of man-hour and also gives rise to interdepartmental rivalry and ill feeling among the fellow scientists. These situations become counter productive to the progress of science and of the institution. It is important to note the observation made in the World Bank report (1991) "Competition between research groups may be viewed as being one of the strength of research in developed countries. However, in developing countries with very limited scientific means, such competition cannot only have a devastating impact when it occurs between small groups, each of which is too small for a sustained qualified research effort."

3.2.1 Fisheries Science Disciplines

On the basis of the mandate of the institute it has to have the technical divisional set up. But before it is discussed in detail it is important to indicate major fisheries disciplines, which will be helpful in deciding the institutional set up. Fisheries and Aquaculture disciplines along with the relevant subjects and certain important topics for higher specializations are broadly indicated.

Fisheries Science

Discipline	Subjects	Higher specialization
Fish Biology	Fish taxonomy	1. Fish taxonomy 2. Crustacean taxonomy 3. Mollusk taxonomy 4. Seaweed taxonomy
	Fish ecology	1. Aquatic biodiversity 2. Riverine fish ecology 3. Estuarine fish ecology 4. Marine fish ecology 5. Coldwater fish ecology
	Fish disease	1. Fish pathology 2. Fish histopathology 3. Fish virology 4. Fish parasitology 5. Fish nutritional diseases 6. Fish disease and diagnostic
	Fish genetics	1. Fish population genetics 2. Fish biochemical genetics 3. Advanced fish genetics research methodology
	Fish physiology	1. Fish bioenergetics 2. Fish digestive physiology 3. Fish osmoregulatory mechanism 4. Fish reproductive physiology
	Fish reproduction	1. Fish reproduction & environment 2. Crustacean reproduction 3. Mollusk reproduction 4. Seaweed propagation
	Fish endocrinology	1. Fish pheromones 2. Fish pituitary physiology

		3. Fish cortico-steroids 4. Peptides hormones
	Fish nutrition	1. Fish feed technology 2. Nutrition of prawn & shrimp 3. Nutrition of different stages of life cycle of important fishes 4. Larval nutrition of carnivorous fish
	Fish microbiology	1. Microbial spoilage dynamics of seaweed 2. Microbial spoilage dynamics of fish 3. Microbial spoilage dynamics of crustacean 4. Microbial fish feed
	Fish behavior	1. Fish migration 2. Fish parental behavioral 3. Fish behavior with environmental variables 4. Fish toxicology
	Fish biochemistry & Fish biotechnology	1. Protein and fatty acid profiling and modification in crustacean, mollusks and weeds. 2. Prostaglandin dynamics 3. Immuno biotechnology 4. Nutritional biotechnology 5. Advanced genetical engineering 6. Control of bio-fouling and bio-corrosion 7. Bio-filters
Inland fisheries	Fish population dynamics	1. Evaluation of different models

	2. Biosphere resource management.
	3. Fish population dynamics in lacustrine condition
	4. Fish population dynamics in fluvialite condition
	5. Migratory fish population dynamics
Aquatic ecology	1. Riverine ecology
	2. Estuarine and brackish-water lagoon ecology.
	3. Reservoir ecology
	4. Pond ecology
	5. Management of inland aquatic ecosystem
	6. Aquatic pollution and its management
	7. Ecorestoration
Aquatic chemistry	1. Running water quality control
	2. Pond water quality control
	3. Soil and water interaction
	4. Micro-nutrient and aquatic productivity
	5. Microbial dynamics
	6. Eutrophication
	7. Nutrient budgeting
Culture based capture Fisheries	1. Cage and pen culture
	2. Rehabilitation of flood plain fisheries
	3. Running water fish culture
Limnology and reservoir fisheries	1. Reservoir fisheries management
	2. Rehabilitation of fish habitat on damming the river

	3. Mitigation of adverse effect in commissioning and decommissioning of dams
Estuarine/lagoon fisheries	1. Estuarine and lagoon fish ranching 2. Island fisheries management 3. Coastal fisheries management 4. Mangrove management 5. Coral reef habitat management
Cold water fisheries and aquaculture	1. Hill stream fish habit and habitat 2. Ranching in coldwater 3. Temperate aquaculture
Aquarium and sport fisheries management	1. Freshwater aquarium 2. Estuarine aquarium 3. Sea water aquarium 4. Sport fish management and conservation
Inland craft and gears	1. Design of energy efficient inland crafts 2. Design, installation and maintenance of modern machinery for net and trap making
Freshwater aquaculture	1. Pond soil and water quality management 2. Food chain and energy dynamics 3. Freshwater fish culture 4. Freshwater shellfish culture 5. Freshwater mollusks culture

		6. Freshwater fish food organism and weed culture 7. Frog culture 8. Pearl culture 9. Intensive and Hi-tech culture system 10. Aquaculture planning and management
	Brackishwater aquaculture	1. Aquaculture planning and management 2. Intensive aquaculture 3. Water and soil quality, management and maintenance 4. Crustacean production 5. Mollusks farming 6. Seaweed onshore production
	Integrated fish farming	1. Different models of integration 2. Paddy fish culture 3. Pond dyke production system
	GIS and inland fisheries	1. Remote sensing for fisheries management 2. Remote sensing of aquatic ecosystem for productivity and pollution status 3. Establishment of GIS for inland fisheries
Marine Fisheries	Marine biology	1. Marine biogeochemcial cycles 2. Community inter relationship of pelagic, intertidal and benthic organisms

	3. Ecology and control of fouling and boring organisms
Marine fish biology and fish behavior	1. Oceanic physiochemical variables and fish migration
	2. Behavior with baits, net and other capture devices
	3. Oceanic mining and fish habitat destruction.
	4. Commercially important fish biology
Fish population dynamics	1. Mathematical models in stock assessment
	2. Stock recruitment analyses
	3. Biochemical population studies
Fisheries oceanography	1. Oceanic currents and fish habitat
	2. Oceanic variables and their effect on coastal and oceanic fisheries
Fishing technology	1. Management and maintenance of modern equipment on board vessel.
	2. GPS technology
	3. Cost effective gears for fishing
Sea farming	1. Seaweed farming
	2. Mollusk farming
	3. Crustacean farming
	4. Cages, ropes and raft design and maintenance

	GIS and marine fisheries	1. Status of potential fishing zones. 2. Fishing forecasting 3. Management of marine protected areas 4. Delineation of marine biosphere reserves and coral reefs 5. Status of fisheries resource
Fisheries Economics & Statistics	Fisheries resource assessment & statistics	1. Frame works for catch monitoring survey 2. Fisheries statistics and dynamics
	Fisheries natural asset accounting	1. Inland resources 2. Marine resources
	Planning & evaluation	1. Different approaches of planning and economic evaluation
	Resources economics	1. Marine fisheries resource economics 2. Estuarine fisheries resource economics 3. Freshwater fisheries resource economics 4. Bioeconomic modeling
	Fisheries cooperative	1. Fisheries cooperatives management
	Finance management	1 Institutional finance and credit 2. Financial system relevant to fishing industry 3. Fish crop insurance
	Fisheries business management	1. Rural business management

		2. International trade
	Fisheries market forecasting and management	1. Market management 2. Fish marketing and price analysis 3. Advertisement
	Socio-economics	1. Gender issues in fisheries 2. Management and administration of social organizations 3. Community development 4. Entrepreneurship development
	Socio-anthropology	1. Fishing community, culture and customs 2. Fishing community migration
Fisheries Extension	Inland fisheries extension	1. Different models and approach to inland fisheries extension
	Aquaculture extension	1. Different models and approaches to aquaculture extension
	Marine fisheries extension	1. Different models and approaches to marine fisheries extension
	Extension methodology	1. Extension education 2. Agricultural extension methodology
	Fisheries communication and journalism	1. Mass communication 2. Journalism
Fisheries Engineering	Gear, crafts & engines	1. Design, fabrication, installation of modern gear and crafts
	Aquaculture structures & implements	1. Cost and energy efficient design of different implement

		2. Farm design in fresh and brackish water
	Electronics	1. Modern electronics appliances and their application in aquatic ecosystem
	Global positioning system	1. Relevance to marine, reservoir and riverine fisheries.
	Fish passes, ladder & other fisheries facilities design and construction	1. Cost effective and energy efficient design and maintenance of fisheries facilities. 2. Rehabilitation of existing facilities
	Eco-restoration	1. Rehabilitation of threatened species 2. Bio leaching of heavy metals from coastal areas 3. Restoration of mangrove 4. Restoration of flood plain fisheries
Fish Processing	Biochemistry and microbiology	1. Irradiation preservation 2. Advanced atmospheric packaging 3. Mycotoxins dynamics 4. Microbial contaminants 5. By product utilization
	Product development and value addition	1. Fibreized products 2. Organic preservation 3. Shelf life enhancement 4. By catch value addition 5. By product utilization 6. Value addition of aquaculture products

	Quality control, and packaging	1. Advances in modern packaging 2. Modern method for monitoring of contaminants in fish crustacean, mollusks and seaweed
	Management of freezing plants, cold storage and ice plants	1. Design of cost effective and energy efficient freezing plants, cold storage and Ice plants
	Storage and transportation	1. Transportation of live fish 2. Transportation of live crustacean 3. Transportation of live mollusks
Fisheries Legislation	Environmental law	1. Quarantine mechanism 2. Legal status of bio diversity 3. International environmental law 4. Environmental impact assessment
	International fisheries legislation	1. Sovereign right of coastal states 2. International commissions on fisheries 3. Responsible fisheries 4. Statutory quality control laws
	Common property aquatic legislation	1. Evolution of aquatic legislation

		2. Interstate relationship for fisheries in common rivers and lakes 3. Common resource equity and sustainability aspects
	Fisheries rules, acts and laws	1. Fisheries acts and laws in developed countries. 2. Fisheries acts and laws for environmental sustainability and for social equity
Fisheries Planning and Management	Research planning and management	1. Planning in national, regional & international organizations. 2. Research management information system.
	Institution building and strengthening	1. Infrastructure development 2. Research infrastructure development 3. Institutional organization
	Capacity building and HRD	1. Human resource planning 2. Institutional capacity enhancement
	Project preparation, monitoring and evaluation	1. Project planning and methods and mechanism for monitoring and evaluation of projects
	Fisheries administration and governance	1. Public administration 2. Fisheries institutes governance 3. Management of fisheries research institutes 4. Fishery corporate, NGO etc. administration and management

3.2.2 Aquaculture Science Disciplines

Aquaculture Science Disciplines with subjects and certain important topics for higher specialization are given below, though some of them are similar to that of fisheries but they need to be mentioned here specifically for clarity :

Aquaculture Science

Discipline	Subjects	Higher specialization
Fish Biology Same as shown under Fisheries Science		
Aquaculture	Freshwater aquaculture	1. Pond water quality management 2. Food chain and energy dynamics 3. Intensive and Hi-tech aquaculture 4. Freshwater fish culture 5. Freshwater shellfish culture 6. Freshwater mollusks culture 7. Freshwater fish food organism and weed culture
	Aquatic ecology	1. Riverine ecology 2. Estuarine and brackishwater lagoon ecology. 3. Reservoir ecology 4. Pond ecology 5. Management of inland aquatic ecosystem 6. Aquatic pollution and its management 7. Ecorestoration
	Aquatic chemistry	1. Running water quality control 2. Pond water quality control 3. Soil and water interaction 4. Micro-nutrient and aquatic productivity

	5. Microbial dynamics 6. Eutrophication 7. Nutrient budgeting
Coldwater aquaculture	1. Hill stream fish habit and habitat. 2. Ranching of cold water fish 3. Temperate aquaculture
Brackishwater aquaculture	1. Intensive Aquaculture 2. Water quantity and quality management and maintenance 3. Crustacean production 4. Mollusks farming 5. Seaweed onshore production
Integrated fish farming	1. Different models of integration 2. Paddy fish culture 3. Pond dyke production system
Farming system research	1. Major crop based FSR 2. Different models of farming system research
GIS and aquaculture	1. Remote sensing for aquaculture management 2. Remote sensing of aquatic productivity and pollution status 3. Establishment of GIS for inland aquaculture
Sea farming	1. Farming of weeds 2. Farming of mollusks 3. Farming of crustaceans 4. Farming of fish

Aquaculture Extension	Aquaculture extension	1. Different models and approaches to aquaculture extension
	Extension methodology	1. Extension education
		2. Agricultural extension methodology
	Aquaculture communication & journalism	1. Mass communication
		2. Journalism
Aquaculture Economics, Socioeconomics & Statistics	Aquaculture statistics & resource assessment	1 Aquaculture statistics and dynamics
	Aquaculture natural asset accounting	1. Inland resources
		2. Marine resources
	Aquaculture resource economics	1. Pond resource economics
		2. Estuarine resource economics
		3. Sectoral conflicts resolution
		4. Bioeconomic modeling
	Cooperative and marketing	1. Aquaculture cooperatives management
		2. Fish marketing and price analysis
	Finance management	1. Institutional finance & credit
		2. Financial system relevant to fishing industry
		3. Fish cop insurance
	Aquaculture business management	1. Rural business management
		2. International trade
	Aquaculture market forecasting and management	1. Market management
		2. Advertisement
	Socio-economics	1. Gender issue in aquaculture

		2. Management and administration of social organizations 3. Community development 4. Entrepreneurship development
	Socio-anthropology	1. Fishing community their customs, tradition, culture and migration
Aquaculture Engineering	Aquaculture structures and implements	1. Cost and energy efficient design of different implement 2. Farm, cage and pen design 3. Modern electronic appliances and their application in aquatic ecosystem
	Remote sensing	1. Remote sensing of aquatic ecosystem
	Global positioning system	1. Relevance to marine, reservoir and coastal ecosystems.
	Geographical information system	1. GIS for open water 2. GIS for controlled water
	Eco-restoration	1. Rehabilitation of habitat for threatened species 2. Bio-leaching of heavy metals from coastal areas 3. Restoration of mangrove 4. Restoration of water structure in flood plain fisheries
Fish Processing	Biochemistry and microbiology	1. Irradiation preservation 2. Advanced atmospheric packaging

		3. Mycotoxins dynamics 4. Microbial contaminants 5. By product utilization 6. Mechanism of spoilage of fish, mollusk, crustacean and seaweeds
	Product development and value addition	1. Fibreized products 2. Organic preservation 3. Shelf life enhancement 4. By catch value addition 5. By-product utilization
	Quality control and packaging	1. Advances in modern packaging 2. Modern method for monitoring of contaminants in fish, crustacean, mollusks and seaweed
	Transportation	1. Transportation of live fish 2. Transportation of live crustacean 3. Transportation of live mollusks
Aquaculture Legislation	Environmental law	1. Quarantine mechanism 2. Legal status of biodiversity 3. International environmental law 4. Environmental impact assessment
	International aquaculture legislation	1. Sovereign right of coastal states 2. International commission on aquaculture

		3. Statutory quality control laws
	Common property aquatic legislation	1. Evolution of aquatic legislation
		2. Interstate relationship for coastal resources
		3. Common resource equity and sustainability aspects
	Aquaculture rules, act and law	1. Aquaculture acts and laws in developed countries
		2. Aquaculture laws and acts for environmental, sustainability and social equity in aquaculture
Aquaculture Planning and Management	Research planning and management	1. Modeling of planning in national, regional and international organizations
		2. Research management information system
	Institution building and strengthening	1. Infrastructure development
		2. Research infrastructure development
		3. Institutional organization
	Capacity building and HRD	1. Human resource Planning
		2. Institutional capacity enhancement
	Project preparation, monitoring and evaluation	1. Project planning and methods and mechanisms for monitoring and evaluation of projects
	Aquaculture administration and governance	1. Public administration
		2. Aquaculture institutes governance

3. Management of aquaculture research institutes
4. Aquaculture corporate, NGO etc. administration and management

3.2.3 Establishment of the Divisions in the Institute

With the above background information institutes should establish its divisions. Generally, fisheries research institutes are established to look after the whole sector but with the growing importance of aquaculture, institutions are being established to exclusively work on aquaculture research. However, both types of the institutions should be mandated to undertake research for biological manipulation for optimizing productivity, ecological interventions for sustaining productivity and viability of resources and economic management for optimizing economic return and bringing social equity. All these require multidisciplinary approach.

Biological manipulations involve disciplines of biology with different subjects such as, Taxonomy, Ecology, Diseases, Genetics, Physiology, Reproduction, Endocrinology, Nutrition, Microbiology, Behavior, Biochemistry, and Biotechnology. Ecological intervention will need the support of Aquatic Ecology, Aquatic Chemistry, Aquaculture Engineering etc. Economic management will be heavily dependent on Economics, Sociology, Marketing and Management Science. At present biological management has occupied almost the entire research thrust in many developing countries. Therefore, the other fields obviously need a major support (Sinha *et al* 1992).

The research needs for the forthcoming decades will vary from region to region and countries to countries but in general higher and optimal productivity of aquatic resources and organisms will remain the primary concern for research. Management of primary productivity and maintenance of water quality will be a dominant component of research, but genetically developed and adopted organisms both plants and animals will remain the basis of future aquaculture in freshwater, brackish water and marine water.

Engineering ingenuity will be desperately needed for better water management devices along with reuse of water and recirculating systems for closed and hi- tech aquaculture systems. More research will be needed to control microorganisms, pathogens and parasites and to fight against diseases.

Development of vaccines and research to improve the immune system of cultivated species will be the high priority research. While, quality control will have to be rigorous, research for post harvest loss and product will be urgently required. Development of mariculture technology will become very important for many developing countries (Sinha 2001). Waste needs to be viewed as an important resource and a technology to use it hygienically for healthy fish production will be needed. Pollution of aquatic resources, overexploitation of fish stock, and conservation of fish biodiversity, environmental sustainability, and effect of weather aberration because of global warming on fish habit and habitat and social equity aspects will need serious research efforts. In addition to those, market management nationally, regionally and globally will demand much needed research support. Keeping these in mind research institutes is to organize their scientific divisions for fisheries and aquaculture research.

Research institutions are normally established according to the basic science disciplines or on the resources or on individual commodity. Fisheries institution established under disciplinal basis should have divisions such as Fish Biology, Inland Fisheries, Marine Fisheries, Fisheries Economics and Statistics, Fisheries Extension, Fisheries Engineering, Fish Processing, Fisheries Legislation, and Fisheries Planning and Management.

Under these divisions relevant laboratories/sections should be established. For example, under Fish Biology division laboratories on Fish Taxonomy, Fish Ecology, Fish Diseases, Fish Genetics, Fish Physiology, Fish Reproduction, Fish Endocrinology, Fish Nutrition, Fish Microbiology, Fish Behavior, Fish Biochemistry and Biotechnology should be set up whereas for Fisheries Economics and Statistics division, sections should be created appropriately to cater the relevant investigations on fisheries statistics, economics and socio-economics. Similarly, institutions on Aquaculture should be set up on the same line.

Sometimes the institutions are established resource-wise with divisions like Freshwater Pond Culture, Riverine Fisheries, Reservoir Fisheries, Lacustrine Fisheries, Brackish water Pond Culture, Marine Fisheries and Mariculture, whereas some institutions are organized on commodity basis such as with divisions like Finfish, Shellfish, Seaweed, and Mollusk etc. Institutions strictly based on resource or commodity wise at times have to have duplication of equipment and manpower.

The most cost effective organizational set up of a Fisheries Institution seems to be the one which has four major research divisions; one on fish biology and other on open-water fisheries, the third on aquaculture and the fourth on fisheries related subjects, supported by one Administrative Department having different sections looking after Administration, Accounts and Finance, Research Management, Information, Library and Documentation, Procurement and Store, Estate Maintenance and Management, and Human Resource Development.

The scientific divisions should be as follow:

Division	
1. Fish Biology	- Fish Taxonomy - Fish Ecology - Fish Diseases, - Fish Genetics - Fish Physiology - Fish Reproduction, - Fish Endocrinology - Fish Nutrition - Fish Microbiology - Fish Behavior - Fish Biochemistry and Biotechnology
2. Openwater Fisheries	- Inland fisheries - Marine fisheries
3. Aquaculture	- Aquatic Ecology - Aquatic Chemistry - Freshwater Aquaculture

	- Culture based Fisheries Management - Brackishwater Aquaculture - Mariculture
4. Fisheries and Aquaculture related subjects	- Fisheries and Aquaculture Engineering - Fish Processing and Product Development - Fisheries and Aquaculture Economics, Socio-economics and Statistics - Fisheries and Aquaculture Extension - Fisheries and Aquaculture Legislation - Fisheries and Aquaculture Planning and Management

Further expansion of the institution needs to be done carefully considering the emergent research issues without duplicating the scientific efforts. At times adding of new division creates difficulty in research management. One of the premier fisheries research institutions in India which has the divisional setup according to the resources such as Riverine and Lacustrine Division, Reservoir Division and Estuarine Division, in course of time added some more divisions such as the divisions of Environmental Monitoring and on Hilsa. Undoubtedly, these two aspects were of great importance. But the major research activities of the existing divisions used to be the detailed ecological and environmental monitoring of all the open water resources. With the setting up of new division of Environment Monitoring, duplication of research efforts between the scientists working in the existing divisions with that of the newly created division resulted. Similarly, creating a Hilsa division resulted in serious duplication and confusion, since all the existing divisions were working on the same fish in different ecosystems.

The often off-shelf prescription of co-operation among the scientists should avoid this situation, but as is well known, symptoms of co-operation are seldom seen in many of the institutions, although they are most desirable. In fact, the situation described would have been resolved not by creating new divisions but giving new orientation of research giving highest consideration on environmental aspects under each division and so on Hilsa. New division should have been created on different river system or types of reservoir or on different estuaries, which possess distinct ecology.

Similarly, research related activities should also be properly grouped under distinct sections set up avoiding the administrative or secretariat model as far as possible. It should be more for facilitating research than controlling research. Research support services departments should be given highest importance to facilitate research.

3.3 Function of the Division

It is important that the institute clearly enunciate the function of the divisions along with the responsibility, accountability and the executive and financial power of each researcher for smooth conduct of research. An outline of the function of different divisions are given below which can be suitably modified by the institutions according to their specific needs and available resources (Sinha, 2000 & 2001).

3.3.1 Fish Biology Division

- To undertake studies on fish taxonomy, biodiversity and conservation of fish genetic resources, fish genetics, biotechnological aspects for upgrading of gene pool, fish reproduction and seed production, fish nutrition and feed formulation, fish disease and diagnostic procedure, fish behavior etc.
- To undertake studies on distribution, migration and environmental preference of important species.

3.3.2 Inland Fisheries Division

- To undertake studies on fisheries resources, inland fisheries and human intervention and mitigating action, stock assessment, population dynamics of commercially important inland fishes, formulation of guidelines for sustainable development.
- To undertake limnological and ecological investigation to assess and monitor productivity of inland water.

- To undertake monitoring of inland aquatic pollution and its impact on fisheries and provide policy guideline on the control of pollution.
- To undertake studies on fishing practices and important gear and crafts used in inland fisheries and evaluate them in terms of their efficacy on the basis of fish behavior and suggest improvement thereof.
- To undertake environmental impact studies of modern development on estuarine ecology and fisheries and to suggest mitigating action and also to provide policy guideline for coastal zone management.
- To study the biological diversity of estuarine, mangrove and coastal ecology and identify the threatened species of fish and prawn.
- To study the detrital flow dynamics and its production potentiality and evolve package of practice to utilize them efficiently.

3.3.3 Marine Fisheries Division

- To conduct research on the biological and ecological production systems that support marine fisheries resources.
- To monitor marine pollution and its impact on the fisheries and also to provide advice and policy recommendations on the control of pollution.
- To conduct studies on resource biomass, catch and fishing efforts and monitor population structure.
- To undertake research on the behavior of important marine species in relation to marine fishing technologies to make improvement thereof.
- To conduct research on the ecology of commercially important marine fish species, including their distribution, migrations and environmental preferences, their modes of reproduction and reproductive seasons, their food and feeding and major predators and to provide advice and policy recommendations on sustainable marine fisheries management.

3.3.4 Aquatic ecology and chemistry Division

- To undertake studies on the basic architecture and production process of freshwater ecosystem in different soil types to help manage them successfully for optimizing fish production.
- To undertake studies on the mechanism for nutrient release from pond sediment to optimize productivity.

- To undertake studies on control of eutrophication and take advantage of higher primary productivity for optimal fish production.
- To develop fertilization schedule for ponds located in different agro-climatic conditions for different methods of aquaculture.

3.3.5. Freshwater Aquaculture Division

- To evolve packages of practices in terms of suitable species, size, stocking density and proportion, feeding, fertilization, harvesting and other management measures for fish culture in freshwater ponds.
- To undertake studies on integrated fish farming, and also on Farming Systems and to evolve different models of such farming systems.
- To conduct environmental impact assessment studies on freshwater aquaculture development.

3.3.6. Culture based fisheries Management Division

- To conduct studies on evolving package of practices in terms of suitable species, size, stocking density and other management measures for culture-based fisheries.
- To undertake study on rehabilitation of inland fisheries.

3.3.7. Brackish water Aquaculture Division

- To evolve packages of practices in terms of suitable species, size, stocking density and proportion, feeding, fertilization, harvesting and other management measures for shrimp and fish culture in brackishwater ponds.
- To undertake studies on integrated fish farming, and also on Farming Systems and to evolve different models of such farming systems.
- To conduct environmental impact assessment studies on brackishwater aquaculture development.

3.3.8. Mariculture Division

- To develop techniques for the commercial culture of potential mariculture species.

- To conduct research on reproductive physiology, genetics, nutrition and environmental requirements of mariculture species.
- To conduct environmental impact studies on mariculture developments.

3.3.9. Fisheries Economics, Statistics and Socio-economics

- To conduct studies on economic, statistic and socio-economic aspects of marine fisheries and make policy recommendations for optimal utilization of fisheries resources by artisanal and industrial fisheries and on allocation of fisheries resources between them to avoid any conflict.
- To conduct socio-economic impact studies of marine fisheries related policies and projects.
- To conduct studies on economic, statistic and socio-economic aspects of inland fisheries and make policy recommendations for optimal utilization of fisheries resources by small scale fishing communities.
- To participate in planning for integrated rural development and resource management.
- To undertake studies on economic and socio-economic aspects of aquaculture, and make policy recommendations on its role in improvement of small-scale fish farmers.
- To undertake studies to evaluate the economics of aquaculture operations in different scales and types of management.
- To undertake socio-economic impact studies of policies and projects related to aquaculture and cultured-based open water fisheries.
- To study the socio-economic impacts of aquaculture technologies on food production, employment, income and foreign exchange earnings.
- To study the social and economic constraints of aquaculture development.

3.3.10. Fisheries Technology and Aquaculture Engineering Division

- To study the fishing technology of open-water and undertake improvement thereof.
- To understand the principles of farm design, water structure and other associated structure for efficient farm management.

- To undertake studies on the engineering aspects of inland aquaculture and mariculture.
- To undertake studies to evolve cost effective hatchery design for fish and shrimps.

3.3.11. Processing, Product Development and Quality Control Division

- To conduct research on the nutritional qualities and the properties of important marine species.
- To conduct research on post-harvest handling, preservation and processing of commercially important species.
- To conduct research on the physical, microbiological and biochemical properties of seafood products.
- To develop product handling, preservation and processing techniques and technology for creating value-added fisheries' products.

3.3.12. Fisheries and Aquaculture Extension Division

- To provide technical support to extension programs relevant to open-water fisheries and aquaculture.
- Develop suitable models for extension activities for open-water fisheries and aquaculture technologies.
- Undertake suitable programs for testing, evaluating and disseminating the research results and technologies.

3.3.13. Fisheries and Aquaculture Legislation Division

- To undertake studies on existing environmental law, fisheries rules, acts and suggest necessary amendments and if necessary propose new ones to deal with emerging constraints and opportunities in the sector.
- To assess the degree of enforcement of fisheries rules and regulations and suggests measures for better implementation.

3.3.14. Fisheries & Aquaculture Research Planning & Management Division

- To undertake studies on existing situation on fisheries administration and governance and suggest improvement thereof.

To undertake studies and develop different models of fisheries institutions strengthening, capacity building and enhancement, HRD and undertake project preparation, monitoring and evaluation.

3.4. Terms of Reference of Staff and their Authority

The term of reference of scientists should be clearly defined to help them to do exactly what is expected out of them. Executive and financial power should be known to them depending on the specific situations and rules and regulations governing the institution. However, it is imperative that the institution considers these aspects as of paramount importance for successful conduct of research and provides the required autonomy, authority, financial and executive powers to its scientists and to the research support staff with strict accountability. Many institutions suffer in absence of such clear-cut guidelines that ultimately affect progress of research. Research studies are location and time specific and any procedural delay in availability of input in time vitiate the whole experimentation.

3.5. Cadre Strength and Disciplinal Support

Scientists are the keys of successful research organizations and every care is needed for recruiting the right type of researcher. In the past, most of the fisheries research organizations were manned primarily by zoologists with specialization in fish biology, or in inland fisheries or in marine fisheries. But for proper evolution of fisheries science and to undertake multidisciplinary approach to solve the field problems, it is important to have specialists from different disciplines as discussed before in section 3.2.1. To circumvent the situation many fisheries institutions started recruiting such specialists from animal science. Though this might be short-term solution but in long-term it may not be sustainable, since there is not much of commonality between habit and habitat of fish and animals or birds as explained in Chapter 1.

The multidisciplinary approach to fisheries research necessitates a much larger base of disciplinal support as discussed in the Section 3.2.1. Accordingly every institution should fix its cadre strength of different levels of scientists and other staff to undertake the research and research related activities. This will also depend on the quantum of financial and manpower resources available.

The requirement both qualitatively and quantitatively will vary over the time. Even so, realistic assumptions need to be made to arrive at such requirements. It is important to have an in-built mechanism to modify these to meet with emerging situations. Fixation of cadre strength is a dynamic process, which needs a review and modification at least every 5-7 years. The essence of this exercise is to have the right kind of people available at the right time, at the right place. Therefore, the institutions should have an appropriate and dynamic human resources planning and development program.

It is also important to understand that research and technology generation is multidisciplinary and a group activity, the success of which is made or marred by the leader of the group. Therefore, right person with leadership and management qualities is the single most important factor of a productive and effective research institution (Grindle 1997). Abidi and Thakur (1998) highlighted some of the skill and professional competence needed in fisheries.

The Indian Council of Agricultural Research in India, which has one of the largest numbers of fisheries research institutions in the world, has done certain useful exercise to fix the cadre strength and identified disciplinal support for its different institutes. The following tables give the details of disciplines and total strength of scientists, senior scientists and principal scientists. While, this needs a serious review in the light of discussion made in the Section 3.2.1, the information will be of some value to other developing countries in undertaking such exercise (Table 2 ➡).

Table 2 **Discipline and Cadre strength of different fisheries institutes under the Indian Council of Agricultural Research**

Discipline	Institute Cadre Strength																															
	CICFRI Barrackpore				CMFRI Cochin				CIFT Cochin				CIFA Bhubaneswar				CIBA Madras				CIFE Mumbai				NBFGR Lucknow				NRCCWF Haldwani			
	S	SS	PS	T	S	SS	PS	T	S	SS	PS	T	S	SS	PS	T	S	SS	PS	T	S	SS	PS	T	S	SS	PS	T	S	SS	PS	T
01. Fish and Fishery Science	24	7	4	35	93	32	18	143	16	2	1	19	15	8	4	27	11	5	2	18	18	3	4	25	6	2	1	9	5	2	2	9
02. Fish Processing Technology	-	-	-	-	1	-	-	1	21	6	3	30	1	-	-	1	-	-	-	-	3	1	-	4	-	-	-	-	-	-	-	-
03. Agricultural Statistics	7	1	1	9	7	1	-	8	2	1	-	3	2	-	-	2	1	-	-	1	3	1	1	5	2	1	-	3	1	-	-	1
04. Agricultural Economics	6	2	-	8	2	-	-	2	2	1	-	3	2	-	-	2	2	-	-	2	3	1	1	5	-	-	-	-	-	-	-	-
05. Agricultural Extension	3	-	-	5	5	3	-	8	5	-	-	5	3	-	-	3	3	1	-	4	4	1	-	5	1	-	-	1	-	-	-	-
06. Agricultural Chemistry	1	-	1	2	1	-	-	1	-	-	-	-	2	1	-	3	3	-	-	3	-	-	-	-	-	-	-	-	-	-	-	-
07. Micro-bioloyg (Animal Science)	4	1	-	5	1	-	-	1	7	2	1	10	2	1	-	3	3	1	-	4	2	1	1	4	-	-	-	-	1	-	-	1
08.Biochemistry (Animal Science)	3	1	-	4	2	-	-	2	9	1	1	11	1	-	-	1	1	1	-	2	3	1	1	5	1	-	-	1	1	-	-	1

Discipline	CICFRI Barrackpore				CMFRI Cochin				CIFT Cochin				CIFA Bhubaneswar				CIBA Madras				CIFE Mumbai				NBFGR Lucknow				NRCCWF Haldwani			
	S	SS	PS	T	S	SS	PS	T	S	SS	PS	T	S	SS	PS	T	S	SS	PS	T	S	SS	PS	T	S	SS	PS	T	S	SS	PS	T
09. Bio-Technology (Animal Science)	3	1	-	4	-	-	-	-	2	1	1	4	2	1	-	3	2	1	-	3	2	1	1	4	3	1	1	5	-	-	-	-
10. Plant Physiology	-	-	-	-	1	-	-	1	-	-	-	-	-	-	-	-	-	-	-	-	-	-	-	-	-	-	-	-	-	-	-	-
11. Veterinary/ fish Pathology	-	-	-	-	5	-	-	5	-	-	-	-	2	1	1	4	1	-	1	2	3	2	-	5	1	1	1	3	1	-	-	1
12. Veterinary / Fish Bacteriology	-	-	-	-	-	-	-	-	-	-	-	-	1	-	-	1	-	-	-	-	-	-	-	-	1	-	-	1	-	-	-	-
13. Veterinary/ Fish Bacteriology & Virology	-	-	-	-	-	-	-	-	-	-	-	-	2	-	-	2	3	-	-	3	-	-	-	-	1	-	-	1	-	-	-	-
14. Veterinary/ Fish Parasitology	-	-	-	-	1	-	-	1	-	-	-	-	2	-	-	2	1	-	-	1	-	-	-	-	-	-	-	-	-	-	-	-
15. Animal/ Fish Genetics & Breeding	2	-	-	2	2	1	-	3	-	-	-	-	4	1	1	6	5	1	-	6	3	2	6	10	3	1	1	5	3	1	-	4
16. Animal/ Fish Nutrition	-	-	-	-	1	1	-	2-	-	-	-	-	3	-	1	4	4	1	1	6	-	3	1	-	4	-	-	-	1	1	-	2
17. Economic Botany	3	-	-	3	1	-	-	1	-	-	-	-	2	-	-	2	-	-	-	-	-	1	1	2	2	-	-	-	-	-	-	-
18. Electronics and Instrumentattion	2	-	-	2	-	-	-	-	2	1	1	4	-	-	-	-	-	-	-	-	-	-	-	-	-	-	-	-	-	-	-	-

4. Issues, Vision and Strategies of Research

Fish exploitation spectrum in different aquatic ecosystems ranges from open-water fisheries to super intensive aquaculture and in between two extremes; there are different fish production systems and methods. Data from various parts of the world when computed for fish production from different ecosystems with different management techniques resulted in many distinct zones. The first zone indicated oceans, rivers and lakes where fish breed and maintain a natural stock, the second one was of lakes and reservoirs where fish were stocked to forage natural food and the third, the smaller lakes and ponds where fish was fed and water was fertilized.

The last zone indicated the super-intensive management where water was re-circulated or aerated and fish was fed with high energy-pellet feed. Intensification of management involving stock/resource manipulation and harvesting results in proportionally higher fish production. As fish production intensifies not only more of seeding and other related requirements are to be met but also studies on ecology, genetics, nutrition, pathology, population structure and reproductive biology become increasingly important for sustaining the quantum of exploitation (Sinha 1990a). However, each system of production faces serious developmental issues and needs research support for their optimal contribution.

Fishing in the pond (Courtesy of Nath and Bhowmick)

Fishing in the Lake (Courtesy of Nath and Bhowmick)

Fishing in the River (Courtesy of Gajbhiye)

Fishing in the Sea (Courtesy of Gajbhiye)

4.1 Open Water Fisheries

Fisheries sector as a whole is facing serious developmental issues, some of them are discussed in general and others are dealt specifically for aquaculture in freshwater, brackishwater and seawater.

4.1.1. Over Exploitation of Marine Fisheries

Marine fisheries face many serious problems. As has been mentioned before, already fishing fleets are 40% larger than oceans can sustain (Linden 2000). Eleven of the world's 15 most important fishing areas have declined in fish population. Also catches of the most commercially valuable species have declined by one-fourth since 1970 (World Watch 2002).

It is evident that much of the inshore water has been heavily over fished. Sometime back, FAO estimated that the cost of over exploitation amounts to about US $ 30,000 million per year and the impact of such exploitation is perhaps the greatest in the developing world (Sinha, 1994).

Decades of over fishing in EU waters have depleted stocks of cod, hake and haddock. It is proposing cuts in national fleets of up to 60 percent over four years. The plan could reduce the total fleet by 8,600 vessels out of a total of 100,000 vessels (EU news - web sites, June 2002).

A view of Fish Landing Center at Alibagh, India (Courtesy of Gajbhiye)

with the UN efforts and a follow up of the UN law of the Sea Convention and the FAO World Fisheries Conference in 1984, an International Conference on Responsible Fishing was organized in Mexico in May 1992. FAO drafted an international code of conduct for Responsible Fishing. The concept of Responsible Fishing involves sustainable utilization of fisheries resources in harmony with the environment and use of capture and aquaculture practices that do not harm ecosystems, resources or food quality.

Marine fisheries research is highly capital oriented. Studies are needed on the standing stock of fish, fish habitat, fish behavior and fishing effort to establish rational management policy and strategy. Sporadic exploratory surveys normally undertaken at national levels do not cover adequately the EEZ area and do not provide sufficient information on the status of both pelagic and demersal resources. Technology for appropriate vessels and gears in developing countries is also lacking for offshore fisheries to exploit under-utilized resources.

Similarly, oceanic processes need to be understood properly to predict the state of the ocean and its reaction to perturbation. Highest consideration needs to be given for waste disposal and its storage in the ocean to avoid any environmental degradation.

Full understanding of the nature and influence of the coastal upwelling leading to high fertility in certain areas that become the sites for major fisheries is essential (Nair, 1998).

Current fishing methods/practices are to be reviewed seriously and upgraded on the basis of the study on behavior of fish. Similarly, existing policy guidelines for conservation and fishery enhancement need to be reviewed and modified to suit the emerging constraints and opportunities.

While, coastal agriculture is facing difficulty with water management, soil salinity and acidity and with other socio economic constraints, high tech fishing fleets from the developed nations are plundering coastal fisheries. In fact, large numbers of families depending on capture fisheries, fish processing and marketing have suffered most. Also marine fisheries development in many developing countries has given rise to economic imbalance and conflict between those engaged in the artisanal and mechanized sectors and between those engaged in coastal aquaculture and fishing sectors (Nair, 1988). This needs to be resolved through implementation of effective policy guidelines based on proper research findings.

4.1.2 Lack of reliable data base

Many countries including China lack reliable fisheries data. It is believed that wrong Chinese data have been masking years of global fisheries decline (Runyan, 2002). He further quoted Pauley who said "our studies showed that whatever leaders set as production targets is what (was) officially reported". Inflating the fish catch statistics is very common in many countries partly because of unreliable methods and means of collection of the data without having any fool proof mechanism or accountability.

Reliable and up to date catch statistics are a necessary prerequisite to stock assessment and fisheries management in open waters. Presently, because of the magnitude of funding involved and lack of strict supervision, the data collection lacks reliability. It is important to review the system and establish database. Establishment of GIS may be helpful. Concurrently, monitoring of the ecosystem in terms of productivity, catch, catchments ecology, landing centers, craft and gears, marketing infrastructure is essential to evolve management policies and evaluate the efficiency of such policies from time to time for further modification.

In order to achieve cost effective catch both artisanal and industrial fisheries would need the assessment of fishery resources based on primary production or chlorophyll distribution or secondary production, trawl surveys or other types of surveys, seasonal stock fluctuating forecasts, precise geographical locations, and data for optimal exploitation rate in terms of number of vessels operated in fishing seasons in different locations for different fish.

A continuous monitoring of the fishery resources particularly of commercial value such as shrimps, lobsters, tunas and cephalopods is necessary with close cooperation between scientist, industry and the fishermen. The information should not only cover the landing of high value catches but also those of low value particularly in terms of their periodicity of availability so that the industry can be able to make best plan of fishing and to convert this low value fishes into value added products.

Similarly, carrying capacity of riverine ecosystem is fast changing and affecting population dynamics and stocks of commercially important riverine fishes. A regular monitoring of the situation is essential for policy formulation for management and sustainable exploitation of riverine fisheries.

4.1.3 Product development and value addition

In general, farmers in the developing countries will suffer because of market distortions caused by subsidies, tariffs and other technical barrier to trade in order to discourage them in being competitive (FAO, 12 Sept. 2003 Cancun, Mexico). This becomes more relevant in case of international trade in fish and fishery product where 80% (by value term) of import is done by the developed countries.

The Hazard Analysis and Critical Control Point (HACCP) become mandatory for all industries producing fish products for USA and EC markets. Use of antibiotics, overexploitation and loss of biodiversity become serious public concern in the developed countries. Recently, EC enforced a law on traceability of fish, which requires producer to indicate commercial and scientific name of the species, its origin, method of production; wild or cultured and the ocean of origin in case if it is marine (FAO, 2002). In addition to these, use of biodegradable and recyclable plastics in packaging is becoming mandatory in many countries (Gopa Kumar, 1998).

Despite all these, the trade of developing countries is progressively getting transformed from export of raw materials for the processing industry of developed countries to value- added products. Individually Quick Frozen (IQF) compared to block frozen items is most favored in effluent markets.

Traditional fish drying on sea-shore in Surat in India (Courtesy of Gajbhiye)

Similarly ready –to cook/ ready to serve type food items are getting popular. Developing countries with considerable number of varieties of dishes and tradition of organic processing with spices and herbs, offer diversified ready to cook dishes. Market research for such products needs to be done continuously to monitor consumers demands. Strict quality control, products free from pathogen and antibiotics, modern packaging and meticulously meeting the export requirements are prerequisite for gaining and retaining international markets.

4.1.4 Onslaught on coastal ecology

During the last four decades, large tracts of mangrove forests were cleared for fuel wood, paddy cultivation, saltpans, human settlement and for shrimp farming in many Southeast Asian countries. Large-scale destruction of mangrove has a number of ecological implications, since it provides natural breeding habitat and acts as nurseries for many marine and freshwater fish and prawn. Apart from soil erosion and subsequent increase in sediment load, the loss of forest cover upsets the natural balance of mangrove ecosystem and adds to the fury of cyclones in the area.

Coastal eco-system is further stressed because of growing use of pesticides, insecticides and herbicide in agriculture, increasing discharge of industrial effluent and human and animal waste.

Industrial effluents polluting the mangrove in India (Courtesy of Gajbhiye)

The city of Mumbai in India with a human population density of 25000/ km2 generates about 2.2×10^6 m^3/d of domestic waste water, considerable amount of which enters the bays and creeks. Similarly, the industrial effluents of about 2.3×10^6 m^3/d also go to a larger extent to the various bays (Qasim, 2002). Varying degrees of environmental degradation occur in estuaries, creeks, bays and localized open shore region of many countries.

Degradation of estuarine water directly affects feeding, breeding and nursery grounds of many marine fin and shellfish. This has adverse effects on marine fisheries (Zingde, 1988).

It is important to assess the damage being done to the coastal ecosystem and to evolve scientific guidelines for its rehabilitation and development. A continuous monitoring to assess the impact of such guidelines is essential to take corrective measures in time.

4.1.5 Open water Fisheries an act of ' Resource Mining'

Open water fisheries resources are supposed to be common property until they are harvested. There has always been a competition among the stakeholders for acquiring the maximum share of the resource without giving any consideration of sustainability, leading invariably to over fishing in such water. However, the newly formed Law of the Sea has changed this in case of marine fisheries at the international level, but resources within the Exclusive Economic Zone of any country now belongs to that country and the fishermen compete within the national boundary.

A serious situation exists in inland sector in many countries where water bodies are leased out for fishing and management to the beneficiary. The system of leasing is based primarily for revenue collection in line with land revenue, which has an in-built psychological incentive for over fishing without any concern of the stock size. It is, in fact, an act of intensive resource mining with serious consequences on biological, economical, social and environmental aspect of fisheries of the water bodies. These aspects need in-depth studies to circumvent the constraints.

4.1.6 Man-made changes in the morphology of rivers and fish biodiversity

Of all the changes taking place in the aquatic resources, rivers have been the most affected. It is interesting to note that more than 40,000 large dams bisect waterways around the world and about 500,000 kilometers of rivers have been dredged and channelized for shipping (Hinrichsen 2003). Water diversion for irrigation, industry and urban use has increased 35 folds in the last 300 years. In this regards, Runyan (2001) correctly remarked that, "Ecologically, rivers are under siege. They are being drained, diverted, polluted, and blocked at a rate that has degraded freshwater ecosystem worldwide with more than half of the world's rivers stopped up by at least one large dam (Over15 meters high)".

Although the world's oldest big dam the Sadd el- Kafara – the Dam of the Pagans was built nearly five thousands years before, which is now crumbling in the Egyptian desert and the other great dam built in 750 B.C. at Mar'b in Yemen, it was mainly in the late eighteenth and early nineteenth centauries when numerous dams were constructed all over the world. Following the trend of damming rivers in the eastern colonies of the British empire, the USA, the then USSR and many other countries dammed their rivers and initiated massive programs for taming of rivers for flood control and intensive agriculture (Ward 2002),

The last two centuries have been rightly termed as an "Age of Dams", with the underlying principal of "water cannot be left to itself". The utility of the dams has been questioned at times and some nations are considering decommissioning of dams whereas some are going ahead of damming large rivers. While, USA, France, Thailand have started decommissioning some of their dams, China and India are going ahead with the construction. Three George Dam recently completed in China is the largest hydroelectric dam in the world and will threaten existence of numerous fish species (Mc Cully, 2001).

Runyan (2001) noted that " in 1998, Bruce Babbitt, the US Secretary of the Interior, travelled across the country to several rivers on a "Sledgehammer tour" – not to break the ground for new construction, but to tear four dams down". Decommissioning of the Mc Pherrin Dam on Buttle creek in California resulted in 20,000 Pacific salmon swimming upstream the following season.

When a river is dammed, its self-purification capacity gets affected. Deforestation and watershed modifications disrupt the protective cover of the soil, resulting in the alteration of the stream flow, increased upstream erosion, turbidity of water, siltation downstream, increased water temperature with low oxygen and wide salinity fluctuations, all of which adversely affect fisheries of the river systems.

A reduction of flood land and the change in the flow of water from fluviatile to lacustrine conditions imposed serious constraints on riverine fish habit and habitat and particularly on their breeding. It affects their natural recruitment in the river systems. However, over the time those fish that survive in changed situation start thriving and those that could not tolerate such change start getting replaced by the former. Thus, fish species composition changes.

The rainy season and freshwater supply directly influence all freshwater fish in their breeding and raising of young ones. There also exists a significant positive relationship between freshwater flow and marine fish production, many marine fish spawn in estuaries or flood plains or undertake reproductive migration at times of peak run off. A decrease in freshwater flow increases salinity and favors marine predatory fish to invade the nursery areas. The effect of the Aswan High Dam on the coastal waters of the Mediterranean is well documented (Drinkwater and Frank 1994). Further, reduction in nutrients transport to the sea reduced catches of *Sardinella* (Shuman, 1995).

It is reported that down stream of the Aswan High Dam in the Lake Nasser only now 25 species of fish are found as compared to 75 fish species in 1940. Similarly, several fish species have been affected in the upstream of the Lake Kariba in Zambia and Zimbabwe. While, *Clarius garepinus, Labeo spp* and *Barbus* species declined, *Protopterus annectens* has become extinct. In Brazil in the Parana River, *Piaractus mesopotamicus* and *Brycon orbignyamus* were eliminated.

The dams on the East River, a tributary of the Pearl River have seriously affected Chinese shad and *Cirrihinus molitorella. The* Chenderoh Dam on the Perak River in Malaysia showed severe effects on Cyprinidae species.

The Danube and the Volga in Europe and the Lower River Murray in Australia are showing decrease in fish biodiversity. In fact, 24 native fish species were lost in the Aral Sea, the fourth largest freshwater lake in central Asia. Same situation exists in Pakistan in the Indus and the Jhelum rivers. In India, of 375 fish species in the Ganges, 46 are endangered, and of 127 fish in the Brahmputra 11 are threatened as reported in 1984.

A global assessment of impacts of the large dams on freshwater fish biodiversity showed 27 % of positive impacts compared to 73% of negative impacts.

It is feared that the decommissioning of the dams now may bring another catastrophe to the established fauna and flora including fish and its habitat.

The dams have already played a significant role in destabilizing riverine ecology and resulted in at least one fifth of the world's freshwater fish endangered or extinct. More so in tropics, since the largest number occurs in the tropics of the global population of 24,600 living species of fish, belonging to 482 families and 57 orders (Nelson, 1994).

Such examples of vanishing fish biodiversity are many more from all over the world. It is ,therefore, important that research attention should be focused on their reproductive biology and controlled propagation and culture so that their conservation could be possible.

Unfortunately, it is not only loss of habitat that has threatened aquatic biodiversity but also haphazard introduction of exotic species resulted in many casualties. For example, the deliberate introduction of Nile perch into Lake Victoria in the late 1950s for increased fish production resulted in the extinction of about 200-300 fish species. Similarly, zebra mussel, a native of Eastern Europe arrived in the Great lakes in 1988, released most likely through the discharge of ballast waters from a cargo ship, started invading rapidly throughout the region. As a result the winged maple leaf clam once found in abundance in the St.Croix River in the upper Mississippi River Basin has now become a threatened species (Hinrichsen, 2003)

It is a pity that the diversity of living organisms amounting to 5 to 10 million species of plants and animals as a result of 3 billion years of evolution involving mutation , recombination and natural selection (Khoshoo, 1986) is meeting such a fate.

4.1.7 Abstraction of water from rivers for agriculture

The abstraction of water from rivers during summer months for crop cultivation puts serious limitation on fish biological activities because at that time many fish physiological requirements need more oxygen, and food and thus greater amount of water.

Abstraction of water, in fact, is threatening the very survival of the ecosystems itself. Glaring example is the Aral Sea, a freshwater ecosystem, which became half in the last forty years with alarming adverse effect on its fisheries. Water of two major rivers the Amu Darya and the Syr Darya that replenish the Aral is diverted to irrigate millions of hectares of cotton cultivation in Uzbekistan. In China heavy demand of water for agriculture has resulted in the dry lake beds on the Gianghan Plains where 1000 lakes existed in 1950 but only 300 remained after three decades. Similarly, the Lake Chad the second largest lake in Africa is in serious ecological trouble. It has shrunk to one tenth of its former size. In 1960 it was 25,000 square kilometers, but now decreased to 2,000 square kilometers. The lake fisheries have more or less collapsed except of the mud fish (Hinrichsen, 2003).

In fact, present system of wasteful irrigation cannot sustain agriculture production itself. It is projected in India that the present per capita per annum availability of 2001 cubic meters of water will be reduced to the stress levels of 1700 cubic meters in the next 2-3 decades and the total availability of water to the agriculture sector will be reduced to about 75% by 2020 from the present share of 89% (Paroda, 2001).Unfortunately it is not that the crop production and industrial production are taking major share of water but livestock farming has also emerged as one of the highest consumers of water. For example, one kg of grain-fed beef needs at least 15 cubic meters of water, and one kg of lamb from a sheep fed on grass needs 10 cubic meters compared to 0.4-3 cubic meters of water for a kg of cereals (Kirby, 2004).

Emerging conflicts for water requirement between fisheries, crop and livestock production need to be resolved with proper formulation of policy guidelines.

4.1.8 Willful Destruction of Juvenile Fish

Indiscriminate catches of young ones of fish and shrimp result in incalculable loss of biodiversity. Hilsa juvenile commonly known as jatka is reported to be heavily fished in Bangladesh and India. About 3000 – 4000 tons of jatka constituting 400-500 million are caught annually in Chandpur area of Meghna river in Bangladesh (Mazid, Personal communication). Over 75 million aquatic organisms mostly fish and shrimp larvae are lost annually during wild collection of 1.5 million post larvae of *Panaeus monodon* for shrimp farming in Bangladesh (Sinha, 1996 a). Similar situation prevails in neighboring West Bengal state of India. India loses 20 billion other fish and shrimp larvae for collecting 400 million wild shrimp for farming annually (Sinha ,1996 b).

In Indonesia young ones hurl explosives and homemade bombs, simply made with a beer or water bottle filled with certain mixture of fertilizers and kerosene and a cheap under-water fuse, on school of fish resulting in massive damage of coral reef and mass killing of fish (Ryan, 2001). According to him, they pose the greatest threat to the world's richest underwater habitat. Damage because of the bombs, sedimentation, cyanide fishing and coral mining have contributed to the degradation of 70% of the Indonesian reefs and have left only 6% in excellent condition. Coral reefs are one of the most productive ecosystems for fish habitat. Their ecorestoration based on scientific studies is of paramount importance.

4.1.9 Pollution of Aquatic Resources

Pollution of open water is a common phenomenon world over. World Water Experts Meeting in Japan in 2001 concluded that more than half of the water in the 90% of lakes and reservoirs of the world is polluted. (Ward2002). It is a most welcome sign that almost every country has taken steps to evolve institutional and legal framework to set standard for pollution of water by industrial or agricultural effluents and make regulations to prevent further pollution. They also undertake surveillance and implement enforcement to maintain proper water quality with varying success. Yet, millions of tons of agrochemical and millions of liters of industrial effluents find their way to the river systems in many countries. Coupled with these, ever increasing sewage without being properly treated get into the aquatic ecosystems with disastrous effects on fish and other biota. Studies conducted in India showed the presence of cadmium, magnesium, nickel, copper, zinc, chromium and lead in almost all estuarine areas of the country.

Arsenic has become a threatening contaminant in freshwater whereas mercury in fish is posing serious problem. It is reported recently that after USA, India is a major consumer of mercury. Potential release of mercury into Indian environment is assessed in between 172–200 tons per year. The residual effect of such chemicals on aquatic systems has not been yet assessed. At times, fish kill is heard but it does not become very effective to arouse unrest or political issues or environmental slogans to force law-enforcing mechanism to take care.

Increased population and production of livestock pose serious environmental problems, since the fecal matter is a major source of nitrogen that is being added to the natural cycle. The present global population of animals is around 45 billion, which is expected to grow to 100 billion in another 20 years.

Nierenberg (2001) estimated that presently 2.5 billion pigs and cattle void more than 80 million tons of waste nitrogen annually compared to 30 million tons by the total human population. If not used properly it will be toxicating the terrestrial and aquatic ecosystems. Scientific studies need to be conducted to mitigate the adverse effect of such pollution on fisheries.

A

B

A and B Show industrial effluents polluting the estuary in India (Courtesy of Gajbhiye)

A

B

A and B Show pollution in Thane creek and Sabarmati estuary in India (Courtesy of Gajbhiye)

A

B

A and B Oil pollution on the Malavan beach, India (Courtesy of Gajbhiye)

A

B

A and B Harbor and Port activities creating localized organic pollution in India (Courtesy of Gajbhiye)

Explosive growth of blue green algae and water hyacinth is seen in aquatic ecosystem almost everywhere. Increasing number of toxic blooms producing algae and their blooms are commonly occurring in many places particularly along the coast of China, India, Japan and Korea. It is suggested that the algae are responding favorably to increased sea surface temperature perhaps because of global warming and also because of nitrogen-laden run off (Nierenberg 2001). It is recently reported that with the current rates of release of nitrogen and phosphorus, there would be a 2.4 to 2.7 fold increase in the next 50 years in nitrogen and phosphorus driven eutrophication of terrestrial, freshwater and near-shore marine ecosystems, seriously degrading biodiversity and fisheries production (Halweil, 2001). This needs mitigating action.

4.1.10. Sedimentation of rivers and reservoirs

It is alarming to note the quantum of sediment load in the major rivers of the world as reflected in the table 3

Table 3: Sedimentation in Selected Major Rivers of the World

River	Country	Annual (Million Tons)
Yellow	China	1,600
Ganges	India	1,455
Amazon	Several	363
Mississippi	United state	300
Irrawaddy	Burma	299
Kosi	India	172
Mekong	Several	170
Nile	Several	111

Source : Brown and Wolf (1984)

Development in the catchment area not only affects the river ecosystem and the fisheries but also affects the lakes and reservoirs' micro-benthic and other micro- communities with far reaching ecological implications. Silt deposition in many reservoirs adversely affects the productivity. Many Indian reservoirs show alarming rate of silt deposition. For example, Bhakra shows about 41 million cubic meters of silt deposition annually, whereas Maithon 7.38, Mayurrakhi 2.47, Nizamsagar 10.76, Panchet 11.76, Tungabhadra 50.48 and Ukai 26.84. Poor management of water in the watershed of Bhakara resulted in salinity and water logging. It is reported that the dam was supposed to have a productive life of 630 years but because of heavy sedimentation, it has been reduced by 200 years.

Sedimentation disrupting riverine ecology in India (Courtesy of Nath and Bhowmick)

Study made by the World Bank showed that Ukai is silting up three times faster than predicted (Ward2002). Similarly, Aswan High Dam in Egypt will take hardly 100 years to be filled up with silt, Mangla in Pakistan 75 years, Ambuklao in Philippines 32 years, Matumbulu and Kisongo in Tanzania 30 and 15 years respectively (Brown and Wolf 1984)

Man-made reservoirs, water storage tanks, small reservoirs and irrigation channels offer opportunities to develop culture- based fisheries, which are yet to be developed in many countries. Most of them need scientific management and particularly artificial stocking. Russian and Chinese reservoirs are yielding larger quantities of fish. Indian reservoirs extending to over 3.1 million ha are hardly producing 10-20 kg/ha/yr. Man-made lakes in Africa lacks proper fish culture management. Therefore, reservoir fisheries management technology needs to be evolved and upgraded with the fast changing ecology of the catchment, sedimentation, and nutrient loading.

Incidentally, it is important to mention that the fish fauna of the particular river system from which the reservoir has originated should strictly guide the selection of species for any fisheries enhancement through extraneous stocking of their seed. The choice should be from among the original fauna of the river rather than implanting from other river systems. The same should be applicable in rehabilitation of flood plain fisheries.

4.1.11. Climatic Change, Global Warming and Fish Habit and Habitat

It is a well-established fact that feeding, breeding and migration of fish are highly influenced directly or indirectly by temperature and many other meteorological factors as well as currents etc. Phenomenon of long term fluctuations in catch has been systematically observed in herring since the sixteenth century. Since then, many more reports dealing with long-term fluctuations in marine capture fisheries have appeared.

The long–term fluctuations of a number of species such as the Japanese and Californian sardine have been shown to have some correlation with climatic indices. Klyashtorin's theories and findings infer possible relationship between annual catches and climate changes. Thus, many attempts have been made to relate fisheries cycles to available long-term climatic variability signals to identify the possible causal mechanisms of fish fluctuations to make long and short term forecasts. However, not much has been analyzed on the effects on fisheries from increased emission of carbon dioxide and other heat trapping gases in the atmosphere as well as from modification of the ozone layer by chlorofluorocarbons : global warming through green house gases.

It is expected that evaporation and precipitation will increase, rainfall pattern will shift, and hurricane and monsoon will be intense. With the increase of UV-B. Penetration of sun burning wavelengths into the shallow and transparent pond water may cause skin lesion in brood stocks and juvenile as has been observed in a number of farms in west Ireland in summer of 1984 (Bullock, A.M. and Cuts, R.P.1985, Mc Ardle and Bullock, 1987). Similarly, observations were made earlier on the effects of solar and artificial UV-B radiation on larvae of anchovy *Engraulus mordax (*Hunter *et al.* 1981*)*. However, studies need to be undertaken on commercially important species to know the degree of damage by increased UV-B on brood stock, their sexual behavior, development, propagation and survival of eggs and larvae to evolve their conservation measures.

The changes in the range of temperature will alter the thermal boundaries that currently limit the distribution of many species. A rise in temperature will increase fish pathogens and will also increase the rate of eutrophication. It is expected that frequencies of substrate shift will be increased affecting the fish habitat. Species and biotic communities' shifts have been observed during long-term climatic fluctuations on the US Pacific Coast (Hubbs, 1948) and also over short period because of the El Nino (Geared 1984, Glynn 1984). In the past 25 years, the welling up of deep water in the eastern Pacific Ocean has decreased, along with its subsequent cooling effects of the ocean and creation of trade winds. This in turn is expected to have more and stronger El Nino events (Ward 2002).

Flash floods, intense monsoons and hurricanes resulting in elevated silt load will choke existing aquatic vegetation and that will affect the ecology and substrate. While floods result in a number of fish eggs and benthic organism dislodged, fine silt deposits in the spawning areas suffocate eggs, embryo and young fish and their food. High turbidity will cause inflammation of the gill and kill young fish. Silt provides ideal substrate for bacterial growth and bacterial infection flares up in the water with heavy silt.

The suspended clay particles are known to reduce feeding rates of some planktivorous fish. Fish feeding visually will be unable to stay in increased concentration of silt but those fish using sensors will move in, affecting fish biodiversity. Freshwater supply, timing of rainfall, and intensity of monsoon have direct effect on spawning of a number of fish and any marked and extreme change in these will influence spawning and availability of seed in nature. Similarly prolonged drought and sudden change in the ionic concentration of water has direct relationship on the spawning and seed production of certain fish (Sinha *et al* 1974, Sinha and Woynarovich 1984).

It is important to identify the vulnerable species likely to be adversely affected under such expected extreme climatic events so that their propagation and seed production technology could be evolved to help retain the biodiversity.

4.1.12 Socio-economic aspects

Till recent past, the objective of open water fisheries management was seen as maximizing the yield, within the limits of natural production, i.e., catch equal to the Maximum Sustainable Yield (MSY). Further realization that fishing has cost, as well as the benefits inherent in the catches, led to the concept that the catch should yield the maximum net benefits. Thus, the "Maximum Economic Yield" of "MEY" became the most desirable management aim. But when catches might as well be considered to have a social value, as fish is a prime source of scarce animal protein, the "Maximum Social Yield" should be the desirable aim. Thus, a holistic approach of management based on biological productivity, economics and equity should control the size and operations of the fisheries that they should generate the maximum net social benefits for the present generation while conserving the resources for posterity.

Appropriate research is needed for development of guidelines for management of open water fisheries with the above objectives.

A proper understanding is needed about the socio-economic conditions of the fishermen for their betterment since most of them are the poorest of poor.

Research is to be conducted to study to improve the distribution systems.

Vision for Open-Water Fisheries Research

The above discussion highlights the R&D issues to help form a vision of research, which may be the following.

In conformity with the national objectives, open water fisheries research should generate comprehensive data and information necessary for guideline formulation and evolve fishing technology and biodiversity conservation methodology for a sustainable fisheries management in consonance with environment, equity and economics. It should also assist in monitoring the impact of such guideline to take corrective measures in time and thus becomes a tool for open water fisheries management.

Strategy for Open-Water Fisheries Research

To achieve the vision, the research strategy should be :

- To provide research support for formulation of guidelines for a sustainable development of open water fisheries resources, through conservation of fisheries physical and biological resources and their optimal utilization.
- The evolution of cost effective fishing technology, methods of reduction of post harvest loss, and value addition and development of quality fishery products for domestic as well as foreign markets.

4.2. Aquaculture

Aquaculture in ponds, small reservoirs and lakes, both in freshwater and brackish water and mariculture have a considerable potential for increased production, but it faces serious challenges. The sector urgently needs various R&D support to achieve its full potentialities.

4.2.1. Lack of data base

Similar to the open water resources, there is a serious lack of reliable information on the aquaculture resources, their progressive use, possible area of expansion, system of culture, and contribution in national economy with an inbuilt system of updating the data. There is an urgent need to establish and access to aquaculture information in every country and continent. FAO has been involved in such exercise, however, much more serious efforts of international agencies are required to establish such database and system with the modern technology of satellite imagery and geographical information system. Very useful study made by Kapetsky (1994) on identification of possible areas for aquaculture in Africa, which needs to be further elaborated and undertaken elsewhere in the world to help in planning for future aquaculture development in different countries.

4.2.2. Water Scarcity & Management

Water management, in aquaculture and particularly in shrimp farming, needs to be progressively based on recycling and reusing water. The demand of water is staggering in shrimp farming, for example, to maintain an ideal condition for shrimp growth, water exchange in the pond is essential which starts with 5% per day in the first month and thereafter of about 10% and about as much as 30% before harvesting the crop. In Taiwan to produce 4-11 tons of shrimps, the water requirement is about 11,000m^3/t and for a production of 12-27 tons of shrimp about 29000-43000m^3/t. Such heavy consumption of water affects the ground water, and causes water table to fall and land subsidence becomes a common phenomenon. Water conservation and management are of vital importance for making shrimp farming environmentally compatible (Sinha, 2002).

The rural ponds are used for many purposes including irrigation and domestic water supply, which interfere with aquaculture actively. Water consumption globally has escalated to unprecedented limits and many parts of the world including much of Africa, northern China, parts of India, Mexico, the Middle East and part of western North America show signs of water stress with water table falling, lakes shrinking and wetland disappearing. Around Beijing, New Delhi, Phoenix, and other mega cities, city-dwellers and farmers compete and claim to the same limited supply. Agricultural, industrial and urban demands have been on rise since 1950 putting pressure on ecological limits (Postel, 1991).

Under the situation, equitable share of water for aquaculture is necessary. Within the framework of integrated resource management, requirement of water for aquaculture needs to be considered in land use planning, coastal zone planning and water use allotment particularly in water conservancy programs. Formulation of legal framework for water rights along with rights to water based resources is necessary to allow access and user rights to aquaculturists to farm fish in water bodies which are by and large under public domain in many countries. Water rights in the western world have been invariably centered around navigation. Ward (2002) noted that it was not until 1973 that Colorado in the USA put in place a law that gives guarantee of water for fish and wildlife.

Recycling and reuse of water will need more capital and energy than many countries can afford. Presently, the energy requirement in fish production in freshwater is meager than other animal production. For example, simple technology of fish farming requires 64-418 kJ of energy to produce 1g.of fish protein, compared to 552 required for egg, 624 for broiler, 779 for pork, 1101 for milk, 2869 for beef and 3424 for lamb (Olah and Sinha 1986). But the energy requirement for intensive aquaculture will be much higher. Therefore, it is important that recycling and reusing of water in aquaculture should be energy efficient particularly for the energy-deficient developing countries.

In this regard, wind power potential needs to be seriously considered particularly for coastal aquaculture. Many countries and particularly India and China have the meteorological potential to greatly increase wind power production. With about 1000 megawatts of capacity, India is already among the five leading wind power nations. Other developing countries are also showing interest in wind power (Renner, 2001).

4.2.3. Pond Improvement and Farm Design

Historically, paddy fields and low lying areas in flood plains and those connected to the estuarine creeks became the cradle of aquaculture where inundation, caused either by monsoon rain or tidal water, brought the natural seed of finfish and shell fish which got trapped after the water receded. That eventually gave rise to the operation of "trapping and holding" of fish seed and raising them to table size and that marked the beginning of aquaculture in many countries and particularly in Asia. Excavation of earth for construction of dwelling units, roads or brickfields contributed to the development of ponds and so does the establishment of religious shrines with ponds for storage of water. Ponds were also excavated for irrigation purposes. In addition to these, geographical phenomenon of shifting of river courses provided opportunities to make rows of ponds by putting embankment to make smaller units resulting in a series of ponds. Such situations are very common in eastern India. Similarly, paddy fields and impoundment near the tidal incursion provided the ground for shrimp and marine finfish culture (Sinha, 1999).

The older ponds are almost in a defunct state in many places. There are many reasons of apathy towards pond improvement. Seldom the pond belongs to an individual, most of the times it has multiple ownership. Individual ownership of the parent over the time got multiplied among the children. A joint decision for investment on pond improvement becomes difficult because of varying interest of the partners. Certain countries adopted policies for state control on such properties, which have not been very successful because of many reasons.

Balarin (1988) while listing the number of ponds, sizes and production figures from 53 countries of Africa indicates that of 270000-321000 freshwater ponds built in these countries, only 25% remain functional and rest are disused and need urgent rehabilitation spanning of about 23000-79000 hectares.

Even after hundreds of years of such traditional aquaculture practices, which improved through many scientific innovations over the years, construction of ponds and farms remain still primitive and lack proper engineering design. Fish crop continues to be seriously affected because of flooding and drought.

The situation is very serious in brackish water aquaculture where in absence of proper water structure, water management becomes very difficult and directly affects the productivity. Coastal aquaculture is badly affected due to lack of proper engineering design particularly for soil and water quality, silt, and effluent management. The situation is aggravated more because of frequent cyclones and mangrove deforestation. The scientific design should provide mangrove belt along the embankment, sedimentation tank and farm effluent treatment setup, having provision for additional ponds for secondary aquaculture, so that the effluent is not wasted or seeped out to other shrimp culture pond.

4.2.4. Pond Productivity and Eutrophication

Many countries in the world are blessed with very high rate of photosynthesis and thus high aquatic productivity. But if is not optimally utilized through fish culture, it gradually disrupts the aquatic environment. Further, with the increased nutrients run off to the aquatic ecosystem, eutrophication sets in, which results in serious infestation of water hyacinth and other obnoxious weeds. The gradual appearance of the aquatic weeds from submerged variety to emergent variety shows highly distressed condition of the ponds. With the advent of the emergent weeds more and more silt is trapped and the quantity, quality and productivity of water are adversely affected.

In India over 60-70 % of ponds and oxbow lakes are infested with water hyacinth (Ayyappan- Personal communication). Such situation exists in many developing countries and water hyacinth has become one of the dominant factors for stressing and disrupting the aquatic ecosystem. Two parents plants of water hyacinth produce about 30 offspring in 23 days resulting in 12000 plants, with a total wet weight of 470t/ha/4months, while registering 5% gain in weight every day (Barrett 1989). In addition to out- growing other

Water Hyacinth Chocking the Pond in India

Enrichment of a Water Body with Birds' Dropping in India (Courtesy of Gajbhiye)

Ox bow lake infested with aquatic weeds in India (Courtesy of Nath and Bhowmick)

aquatic plants and animals, its growth results in trapping silt in large quantities and the loss of water through transpiration and plant growth, which adversely affects the carrying capacity of the ecosystem. Since water hyacinth contains 95-97 % of water, it amounts to a potential loss of 446 tons of water/ha in 4 months, which is certainly most undesirable.

The situation calls for development of technology for productivity management along with soil and water quality maintenance, and reducing the health hazard to the fish, animal and human being that consume pond water. Undertaking aquaculture properly in such water bodies utilizes the natural productivity and thus reduces the chance of disrupting the aquatic environment. Fish is a major ecological player and aquaculture needs to be viewed as a tool to utilize primary productivity to help maintain water and soil quality in the pond.

There is a growing awareness all over the world of the need to recycle sewage waste into protein wealth. Sewage water waste provides scope for large-scale culture of protein rich algae at the primary level and fish at the secondary treated level, which improves water quality, besides producing fish for food or fish meal for animal feed.

Management and recycling of the nutrients from pond humus is very important, since it is a potential manure with long lasting fertility effect. One ton of pond humus has fertilizing effect equivalent of 6 kg of ammonium sulfate. It is also estimated that a properly managed pond produces over 50 tons of humus and silt/ha/yr. Underutilization of humus results in deterioration of condition of the pond soil and consequently anaerobic state sets in the pond bottom affecting adversely the productivity.

Similarly, artificial shrimp feed and fecal matters of shrimp also contribute to degrade the pond condition in semi-intensive and intensive culture. The effluents from such farms contain heavy load of organic materials. It is estimated that one ton of shrimp production results in loading of 57.3-118.1 kg of N and 13.0-24.4 kg of P in the system. There is a serious need of improved management technique. Kaushik (2002) suggested that fish meal component of artificial feed contributes to significant release of phosphorous into the environment. Though the availability of phosphorous from fish meal is variable, efforts are being made for replacement of fish meal by plant products.

4.2.5. Fish Disease

Fish disease epidemics are on increase and will continue to be a threat to aquaculture development. The last two decades have seen the devastation caused by the Epizootic Ulcerative Disease Syndrome (EUS) in carp, catfish and many other fishes in wild and under culture conditions in different south Asian and Australian regions. Similarly, the white spot disease has created havoc in coastal aquaculture particularly in shrimp farming all over Asia. Recently, Asian Development Bank and Network of Aquaculture Centers in Asia- Pacific (1998) prepared a valuable report on aquaculture sustainability and environment. It indicated that the disease problems are closely related to water quality and other environmental problems and showed a total loss of $104.6 million /yr. in carp culture in the region which is about half of that in shrimp farming. Globally, the loss is estimated to be over $3 billion because of disease in shrimp aquaculture. Thorough knowledge of identification of pathogen, their mode of infestation, role of environment and of holistic pathology is essential to evolve methods of control of diseases. Health of fish and shrimps becomes a serious issue not only for production point of view but it has serious significance for public health as well as for export. Most of the developing countries have invested substantially in aquaculture infrastructure and thus it is imperative that they should gain and retain export markets (Mazid & Sinha 2003).

4.2.6. Socio-economic Constraints

Of late it has been realized that participation of women is of paramount importance in poverty alleviation and food security, In order to strengthen their involvement, development agencies and the national governments are striving hard to find ways and means to encourage women active participation in such developmental efforts. Women world over have been engaged not only in raising family but also maintaining livestock, net making, processing and selling fish and such other rural activities in and around their dwelling units and play very crucial role for family food security.

Women are in forefront of food-related activities but not that much in actual food or fish production or in procurement from the open water. It is interesting to note that 78 percent of economically active women in India are engaged in agriculture and allied fields. However, aquaculture offers scope for involving women in fish seed production, raising fingerlings, backyard hatchery for prawn, aquarium setting, ornamental fish production, pearl culture and many other related activities. With their child raising intuitive and ingenious ability with great patience and doing fine handicrafts, they are bound to be most successful in above aquaculture activities. However, studies need to be undertaken to evolve ways and means to encourage and facilitate their involvement.

Presently, most of the benefits go to the large fish farmers and entrepreneurs who are generally city dwellers, often alien to the area. They get the leased land of small farmers to operate shrimp aquaculture enterprises. This deprives small farmers of the benefit of coastal aquaculture and leads to social tension. Proper studies need to be undertaken for policy formulation for government intervention.

The conversion of paddy land to aquaculture causes conflicts over land tenure and user rights in certain areas. This needs in-depth study for policy formulation for government intervention to avoid sectoral conflicts.

The price spread from farmers to consumers is at least about three times. It is important to undertake study to reduce the marketing cost so that producers get more profit and consumers pay less.

Lack of firm socio-economic indicators, to assess the progressive socio-economic upliftment of rural poor with adoption of aquaculture is a great impediment in convincing planners and policy makers about the quantified role of aquaculture to get their required support.

There is an acute shortage of skilled fish farmers in many countries where there was no tradition of aquaculture. Study needs to be undertaken to assess the farmers' training needs, their present skill and aptitude to embark on their training in aquaculture activities.

Studies need to be conducted on consumer acceptance of fish and fisheries products to achieve an accelerated rate of growth in trade.

Infrastructure and credit facilities in many countries are far from satisfactory or yet to be properly evolved. There is a strong need to create market infrastructure along with supply centers, and better credit facilities to avoid middlemen. Marketing cost of aquaculture commodity is very high, which needs to be minimized, through proper investment in improved transportation, infrastructure and marketing facilities. Also creation and expansion of small credit and saving institutions are necessary to facilitate the development of aquaculture trading, transportation and processing enterprises. Proper studies need to be undertaken to formulate the policy guidelines in these respects.

4.2.7. Research need for Mariculture

Sea farming or Mariculture i.e. aquaculture of marine fish, crustaceans, mollusks, seaweed etc is another promising area to supplement food production. Marine fish , shrimps, lobster, oysters, mussels, clams, scallops, abalone and seaweed are cultivated in many countries. Already production technology of many of these are either existing or at different stages of development in different countries. However, documentation and evaluation of genetic resources and seed prospecting of cultivable mollusks and seaweed are essential. Also research is urgently needed on predator and disease control, depuration and post-harvest technology, product development, market potential and socio economic impact on fishing communities (Sinha 1990).

Vision for Aquaculture Research

Some of the R & D issues of aquaculture have been highlighted, which may vary in different countries and situations. But in general the vision of aquaculture research should be as follows:

In conformity with the national objectives, evolve technology for conservation and up-gradation of gene pool of fish and develop environmentally compatible, socio-culturally acceptable and economically viable aquaculture technologies for integrated rural and coastal development.

Strategy for Aquaculture Research

To achieve this vision, the major strategies for aquaculture research should be :

- Productivity management and maintenance of soil and water to reduce the risk of public heath hazard and fish disease.
- Stock improvement and feed formulation of commercially important fishes and conservation of endangered species.
- Development of sustainable aquaculture for small reservoirs and culture based capture fisheries.
- Assessment of investment for market infrastructure development, marketing management ,involvement of women and on risks management for a sustainable aquaculture development.

5. Research Formulation and Prioritization

5.1. Research Project Identification

The identification and the formulation of research project are the initial important steps of research; the former helps in finding out the right research study whereas the latter helps to guide the ways to do it rightly. A general framework of research project identification and formulation is shown in flow chart 1.

Research identification process differs widely in different institutions. Selection of the topics of research at the universities leading to post graduate degrees, by and large depends upon the expertise of the available supervisors. In fact, supervisors of the students select the topics. In many research institutions individual scientist under the guidance of the seniors identifies the area and topic of research. The procedure of approval of the project proposal is highly variable in different institutions but basically it needs to satisfy the existing academic, administrative, technical and financial set up.

5.1.1. Data on Background Information

The following background information is necessary to assess the present status of fisheries in the country. This helps in the process of identification of research project.

- Present and potential fisheries resources, existing capture and culture fisheries systems.
- Patterns and levels of fisheries development.
- Total fish requirements for domestic consumption and export.
- Present form and level of fisheries industry.
- Available supporting services i.e., both technical and non-technical,

material inputs, markets and marketing arrangements, storage, preservation/processing, and transport facilities.

- Financial and credit facilities.

An understanding of the present status of industry and the need of the specific research support for its development are essential. An analysis is undertaken to assess the nature of constraint, which is to be removed through proper scientific studies or in other words to find out the malady for remedy.

5.1.2. Constraint analysis

Developmental and research issues confronting open water fisheries management and aquaculture along with research strategies and vision have been discussed in Chapter-4. With clear background information on those aspects and on the present status of the industry, scientist needs to undertake constraint analysis to identify the specific problem, which needs investigation.

An estimate needs to be made of loss in terms of yield, resource base including material and manpower due to each constraint, and expected gains resulting from successful research. This is a difficult exercise but a group of scientists should discuss and decide which type of research studies would be useful to circumvent a particular constraint most effectively. For example, to control the productivity loss whether research should focus on genetic improvement of a particular fish or productivity management of the pond or both. Joshi and Bantilan (1997) suggested similar exercise for identification of a particular research strategy.

An example of constraint analysis, regarding the existing yield gap between the production obtained by the farmers and those that have been possible in aquaculture experimentation is shown in the following table. In general, there may be three categories of constraints for investigation; (i) Production constraints; (ii) Ecological constraints (iii) Socio-economic constraints.

The production constraint may be because of unavailability of quality seed, feed or fertilizers or because of defective design of pond or implement. The ecological constraint may vary from natural calamity to eutrophication, salinity, alkalinity and acidity of pond soil. The socio-economic constraint in the field involves the skill and aptitude of the farmers. In addition to this, legal and political issues, and availability of credit etc. pose serious constraint.

Possible Research Studies to Solve Different Types of Constraints

Actual Yield in Farmers' Pond	**Experimental and Potential Yield**
Production Constraint	**Research Studies**
1. Quality Fish Seed.	· Standardization of quality seed. · Broodstock management. · Quality seed production. · Hatchery management.
2. Fish Feed and Fertilizers.	· Survey of feed ingredients. · Nutritional requirement of fish. · Study of nutrient status of soil. · Aquatic productivity management. · Role of micronutrients in augmenting productivity.
3. Pond and Farm Design	· Control of seepage · Water management studies for upland ponds and low land ponds. · Designing of low cost farms.
4. Implements (Boat, Net etc.)	· Gear and crafts designing for effective catch.
Ecological Constraints	**Research Studies**
1. Flood, Cyclone, Extreme weather condition.	· Studies on vulnerability of fish species under such stress.
2. Eutrophication, Siltation and Weed infestation	· Primary productivity management. · Effect of silt on fish biology. · Weed management.

Actual Yield in Farmers' Pond	Experimental and Potential Yield
Production Constraint	**Research Studies**
3. Salinity, alkalinity and acidity of soil	· Productivity management of problematic soils through fish culture.
4. Other environmental issues	· UV-radiation effects on fish seed raising and brood maintenance.
Socio-economic Constraints	**Research Studies**
1. Skill and attitude of the fishermen and entrepreneurs	· Bench March survey and specific research studies to assess and take corrective measure
2. Legal issues	· Legal issues-policy research
3. Political issues	· Political issues- policy research
4. Credit	· Credit-policy research

Flowchart 1: General Framework of Research Project Identification and Formulation.

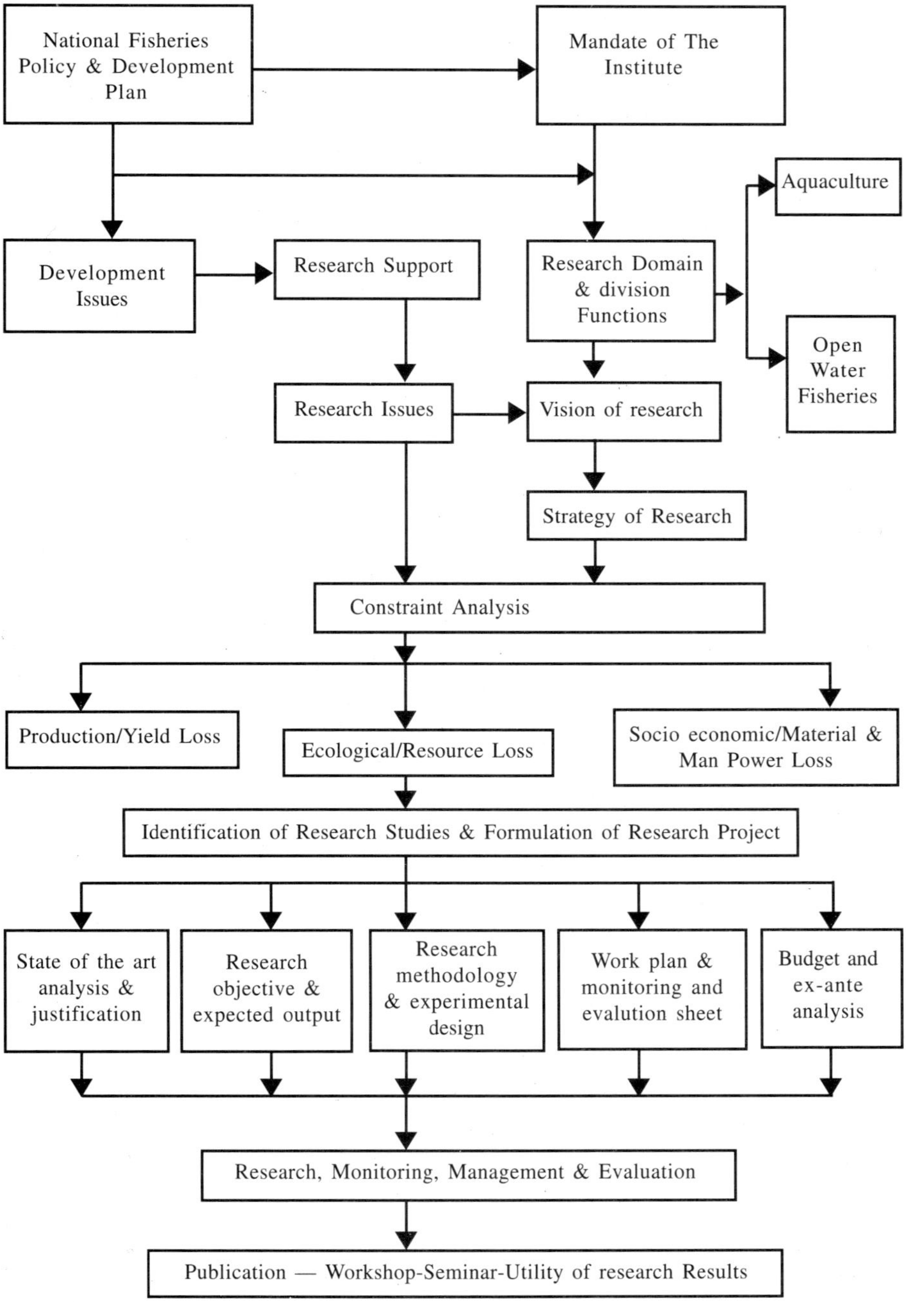

5.1.3 Literature Search

The generation of the new information is as important as the knowledge of the existing information for any meaningful and cost effective research. Thus, the most important initial exercise during formulation of the research project proposal is the literature search and to review the present knowledge available in the field. As Tacon (1992) very aptly said "A prerequisite to the planning and implementation of a research and development activity is first to be aware of the previous studies conducted in this subject area by other related field projects, and by so doing avoiding unnecessary duplication of research effort and the re-invention of the wheel". To add to this, it is highly desirable that the researchers should not be over enthusiastic to become the inventor for all what he/she does, either deliberately or ignorantly about the previous work. These reflect on their scientific integrity and vitiate the scientific culture and tradition.

The literature search has become simpler now with the Internet facilities. However, the Fisheries Branch Library (FBL), which is a part of the David Lubin Memorial Library, the central library of Food & Agriculture Organization of the United Nations (FAO) provides commendable assistance in literature search. It collects publication on all aspects of fisheries from statistics and economics, through legislative and sociological aspects, to the natural and applied sciences of fisheries biology, aquaculture, fish technology and marketing. The extensive collection which has been built up over the past 25 years presently consists of 1600 periodical titles, over 10,000 books, approximately 12,000 FAO reports and several thousands reports of other fisheries agencies.

The Fisheries Department of the FAO produces some 75 publications a year in the following regular series, which are widely distributed:

FAO Fisheries Technical Papers
FAO Fisheries Reports
FAO Fisheries Circular
FAO Fisheries Synopses

FAO Yearbook of Fishery Statistics

In addition to the fisheries publications produced at the headquarters at Rome there is a wealth of documentation produced by FAO fisheries projects in Africa, Asia and the Pacific, Latin America and the Caribbean, the Near East, Eastern Europe and the Mediterranean.

FBL also collects and microfiches FAO publications and documents and maintains computerized database to manage this collection and facilities access. The microfiches of documents are available on request (Collins 1993).

Other relevant rich resources are ICLARM Library and Holdings, Libraries of Freshwater Biological Association, Windermere and University of Stirling in the U.K. and Auburn University in the USA. However, there are many more sources, which should be explored to search literatures relevant to the study.

5.2 Research Project Formulation

5.2.1 Selection of Parameters and Methods of Experimentation

The research plan needs to incorporate the methods of experimentation. There are a number of standard books available for consultation on methodology. It is desirable that the researchers take a very judicious decision in selecting important parameters and the frequency of sampling as well as the site of sampling. The methodology should involve as far as possible simple instrumentation, little working time and labor, but should offer sufficient scientific information. The task is difficult but is crucial as is evident from the foregoing discussion citing an example of assessing nutrient status in confined water and in open water.

Both aquaculture and open water fisheries management research involves studying the soil and water qualities. To undertake such investigation in case of a pond, Olah and Sinha (1984) included 31 parameters; most of them could be investigated without much of sophisticated equipment and specialized skill, and yet provide enough information for judicious management. The selected parameters are:

1. Pond code
2. Water area(ha)
3. Age (year)
4. Management
5. Visual color and temperature
6. Transparency (cm)
7. Water depth (cm)
8. Soft sediment depth (cm)
9. Solid sediment depth (cm)
10. Sediment gases ($dm^{-3}.m^{-2}$)
11. Sediment organic-C ($mg.g^{-1}$)
12. Sediment detritus ($g.m^{-2}$)

13. pH
14. Alkalinity (mg.dm^{-2})
15. NH_4-N (µg.dm^{-3})
16. NH_3-N (µg.dm^{-3})
17. PO_4-P (µg.dm^{-3})
18. Dawn Oxygen(µg.dm-$^{3)}$
19. Bacterioplankton (100 cm^{-3})
20. Phytoplankton (cell.dm^{-3})
21. Seston detritus, particles (cm^{-3})
22. Seston (>60 µm, wet weight)
23. Dominant species (>60 µm)
24. Seston (>150 µm, wet weight)
25. Dominant species (>150 µm)
26. Macrozoobenthos (>400 µm.m-^{2})
27. Dominant species (>400 µm)
28. Macrotecton (>400 µm. m^{-2})
29. Dominant species (>400 µm)
30. Macrophyte cover (%)
31. Dominant species of macrophyte

The parameters have been elaborated here to understand their significance:

Age The age of the ponds is important since it has a direct relevance with the productivity of the pond. Generally new ponds are not that productive nor the old ponds when they became eutrophic.

Management Method of management of the pond helps to assess some quantified data on the organic carbon and biogenic nutrient load.

Color and temperature of the pond water The visual color of the pond water gives a simple but important reflection of the basic production processes. Temperature is an important indicator for all physiological, chemical and biochemical reactions and also for solubility of the oxygen in the water.

Water transparency and depth of water It reflects the trophic or production level of the water and is measured with Secchi disc. However, the transparency gets affected by high turbidity or by dense algal population. Depth of water is important to know the total volume of water and it is also important to know how far the sunlight goes deep down into the water.

Sediment depth There are two layers of sediment present at the bottom of pond, one is the soft and the other is hard and solid. The thickness of the layer can be measured with a six to eight meter long bamboo pole provided with a sharp end. The total thickness of the soft plus solid sediment layer has a direct relation to the age of fishponds. Thick sediment having rich nutrients influences the productivity of the pond.

Sediment gases The bottom sediment contains several gases, mainly methane, nitrogen, hydrogen, carbon dioxide and nitrous oxide. Methane and hydrogen are end products of the anaerobic decomposition of organic carbon, whereas carbon dioxide is formed primarily from aerobic decomposition. Nitrogen and nitrous oxide are produced during the process of denitrification. Quantification of this parameter can be done by a standard sediment stirrer grid superimposed by a funnel mounting connected to a gas collector through a rubber tube. This small device can be attached for gas collection to the bamboo pole used for depth measurements.

Sediment organic carbon The sediment layer is formed with the accumulation and deposition of silt and the energy-rich organic matter. The Hargrove sampler is the most suitable tool to collect sediment samples and the sample's ignition procedure is sufficient to measure the sediment organic carbon.

Sediment detritus It consists mainly of decomposing aquatic plants and animals. Organic matter originating from the autochthonous and allochthonous sources is also present in the detritus in pond. The nutritive value of the detritus changes during the different stages of the decomposition. To measure the quantity of this nutrient resource, a simple sieving procedure is adopted with a suitable mesh size. The wet residue is mainly inorganic particles having higher specific weight and stay in the corners of the washing tray during decantation.

Detrital food chain is the other pathway that contributes to the production in an aquatic environment. The rate of decomposition of organic substances varies according to the environmental conditions and the constituents. The cycling of carbon is influenced to a high degree by the oxygen content of water.

All habitats depend on micro-organisms including bacteria and cyano bacteria for nitrogen fixation. Similarly, microorganisms play an important role in the mobilization and immobilization of phosphorus. Solubilization of insoluble inorganic compounds is brought about by some species of bacteria. The mineralization of organic phosphorus compounds is effected through the enzymes (mainly phosphatases) produced by bacteria, algae and even zooplanktons.

Chemical characterization of water column The water is chemically characterized by pH and alkalinity, NH_4-N, NO_3-N and PO_4-P. Measurements of these are done with standard methods. Normally, the pH and alkalinity do not change from pond to pond on the same type of maternal soil. The measurement of NH_4-N, NO_3^-N and PO_4-P indicates the basic inorganic nutrient status of the pond.

Dawn oxygen Generally, there is a high fluctuation in the oxygen content during the day and night. The diurnal oxygen fluctuation is measured to calculate the community metabolism of the whole pond and to quantify the production and respiration process in the ecosystem. As it is well known that many environmental parameters influence the concentration of the oxygen both in time and in space, a single measurement at the end of the dark period would be an important indicator of the risk of fish kill. The oxygen measurement can be done by oxygen electrode in situ.

Bacterioplankton Bacteria are a dominant component of the ecosystem and play crucial roles in a number of important process in the energy flow and mineral cycling, besides serving as food item for fish and other aquatic animals. Generally, the bacterial colonies appearing on the agar surface or in the agar column after a given incubation period, are counted. But since bacteria growing in any kind of nutrient media represents only a very small portion of the total bacterial community present in natural waters it is better to take 100 cm^{-3} pond water preserved immediately in 2% formalin and then filtered through a Millipore membrane filter with a pore size of 0.2 μm . The residue of the filter is stained with erythrosine dissolved in freshly distilled phenol. After drying the membrane filter, the bacteria may be counted under microscope with a magnification above 1200 and make quantitative estimate of main groups taking part in mineralization process.

Seston size fractions The fish food organisms in the water have much shorter generation time than those in the sediment. The generation time of phytoplankton is about 10-20 hours and for zooplankton around 1-4 days, whereas for benthic animals, it is much longer. A rough production estimate can be made on the basis of the biomass of the planktonic organisms and their generation periods. Olah and Sinha (1984) recommended that two seston samples using first, a mesh size of around 60 μm to collect phytoplankton, zooplankton and detritus, and then the second, with a mesh size of around 150 μm to collect larger organisms would be sufficient for quantification of natural food resources available in the water. After centrifugation, the volume of seston may be assessed and the wet weight may be taken. Assessment of the real nutritive value of the planktonic organisms becomes difficult since many undesirable and indigestible algae may be the major component of the fractions. However, formalin preserved samples should be analyzed to have some idea of dominant species of both phyto and zooplankton.

Zoobenthos and zootecton Generally, animals living in the sediment or among the macrophytes have much longer generation time than the planktonic ones, yet they occupy an important place in the natural fish food resources. Ponds with thick anaerobic sediment layer are not favorable to maintain any significant animal populations. Chironomids and Oligochaete species that are fully adapted to the sediment environment find difficulty in establishing themselves in the absence of oxygen and the presence of different reduced compounds in high concentration. Their absence thus, will indicate highly adverse conditions for pond productivity.

Presence of macrophyte provides cover for many large size animals, such as crustaceans, insects and mollusks and thus, they establish and flourish. Benthic animals and the zoobiotecton living among the macrophytes can be identified and counted after a rapid sieving with a mesh size of around 400 µm, and carefully washed in a large white tray for quantitative and qualitative analyses.

Besides, it is also important to visually estimate the quantity of the macrophyte, since it directly affects the availability of plankton. Both the groups compete not only for nutrients but macrophytes obstruct the incidence of sunlight in the water and thus, adversely affect plankton production.

Running Water Quality Assessment

An assessment of running water quality is normally based on the same basic information as discussed before. But the influence of catchment, including anthropogenic pressure becomes highly relevant in running water. Assessments of transparency, pH, DO, alkalinity, nitrate, phosphate, conductivity and CO^2 are required to know the water quality. Assessment of suspended solids, BOD, chloride, phosphate and nitrate are essential to know the quantum of organic load and the degree of eutrophication. Evaluation of metals like Zn, Pb, Hg, As, Cd, and Cr becomes important if pollution aspects are investigated.

Similarly, data on open water fisheries are essential to know the biology of fish species, population structure, yield estimation, and information on craft and gears.

Biological Score System

The National River Water Quality Survey, U.K., designed in 1980 a scoring system for assessment of biological quality of rivers, which was further, modified (Andrew, 1991). Water Sampling sites in the river are scored in the prescribed manner and better water quality shows higher biological score based on the following table:

i) All families of species present are listed

ii) Scores are ascribed to each of these families according to the values indicated in the following table.

iii) Scores for all families are added to give the total cumulative site scores, which indicate the biological quality of rivers as based on the recommendations of the Biological Monitoring Working Party (BMWP). Score is directly proportional to the quality of the water. Values well in excess of 200 indicate pure and unpolluted water.

iv) Sampling sites, method of sampling as well as season of sampling need to be judiciously decided. A minimum frequency of one sampling site per 20-50 kilometers of river is recommended.

Suggested Biological Score System for Different Rivers

E	*Siphlonuridae, Heptageniidae, Leptophlebiidae, Ephemerellidae,*	10
	Potamanathidae, Ephemeridae.	10
PI	*Taeniopterygidae, Leuctridae, Capniidae, Perlodidae, Perlidae,*	10
	Chloroperlidae, Aphelocheiridae.	10
He	*Phryganeidae, Molannidae, Beraeidae, Odontoceridae,*	
	Leptoceridae.	10
T	*Goeridae, Lepidostomatidae, Brachycentridae, Sericostomatidae.*	10
Od	*Lestidae, Agriidae, Gomphidae, Cordulegasteridae, Aeshnidae,*	8
	Corduliidae, Libellulidae.	8
T	*Psychomyiidae, Philopotamidae.*	8
E	*Caenidae.*	7
PI	*Nemouridae.*	7

T	*Rhyacophilidae, Polycentropodidae, Limnephilidae.*	7
M T C Po Od	*Neritidae, Viviparidae, Ancylidae, Unionidae.* *Hydroptilidae.* *Corophiidae, Gammaridae, Paleamonidae.* *Nereidae, Nephthyidae.* *Platycnemididae, Coenagriidae.*	6 6 6 6 6
He Co T D P E Me H	*Mesovelidae, Hydrometridae, Gerridae, Nepidae, Naucoridae, Notonectidae, Pleidae, Corixidae.* *Haliplidae, Hygrobiidae, Dytiscidae, Gyrinidae, Hydrophilidae,* *Helodidae, Dryopidae, Elminthidae, Chrysomelidae, Curculionidae.* *Hydrosychidae.* *Tipulidae, Simuliidae.* *Planariidae, Dendrocoelidae.* *Baetidae.* *Sialidae.* *Piscicolidae.*	5 5 5 5 5 5 4 4 4
M H C	*Valvatidae, Hydrobiidae, Lymnaeidae, Physidae,* *Planorbidae, Sphaeriidae.* *Glossiphoniidae, Hirudidae, Erpobdellidae.* *Asellidae.*	3 3 3 3
Ch O	*Chironomidae.* Oligochaeta (whole class).	2 1

E=Ephemeroptera Pl=Plecoptera He=Hemiptera T=Trichoptra Od=Odonata M=Mollusca C=Crustacea Po=Polychaeta P=Platyhelminthes D=Diptera Me=Megaloptera H=Hirdinea O=Oligochaeta Co=Coleoptera Ch=Chironomidae

Biological Indicators in Water Quality Monitoring

Most of the species have specific requirements for oxygen and nutrients. Availability of a species in healthy state in a particular habitat indicates that its physico-chemical parameter is within the tolerance limits of that species and thus, it acts as a biological indicator. Some species could manage to exist in a difficult environment, but suffer with physiological stress in growth and other biological activities, and thus, act as a 'bio-sensor' for that pollutant or stressor. Some organisms also accumulate toxic pollutant in their tissues which reflect the presence of those substances in the environment and thus act as 'bio-accumulators'.

Qualitative and quantitative assessments of bacteria are used for detecting contamination of water or health risks, but its use as index for monitoring purposes is limited and so is the role of protozoa, algae, and macrophytes. However, macro-invertebrates, particularly those that are sedentary, are very useful in detecting the precise location of pollutants sources, and those having long life histories provide opportunity for examining temporal changes and reflecting the effects of prolonged exposure to pollutant. Thus they are used in evolving pollution indices and diversity indices to show the relative state of water quality (Vaas, Personal communication). The benthic invertebrates reflect the stress levels more conclusively. Therefore, a regular monitoring of the benthic communities of the river at various sites should be done through a uniform and standardized methodology, so that comparative data are obtained for an assessment of the degree of deterioration of the water quality, which has a direct relationship with fisheries.

Fish Stock Assessment and Population Dynamics Studies

The fish stock assessment studies need adequate data on size composition (length or weight) of catch for each stock under investigation. For this, it is necessary to sample the commercial catches according to a pre-determined scheme of periodicity and sites of sampling. The sampling scheme should take into account the area of distribution of the stock and also different types of craft and gears used. Stratified multistage random sampling can be undertaken in the ecosystem suitably which can be divided into several zones on the basis of geographical consideration and fishing practices.

The landing centers are randomly selected in the zones for recording fish landings. Periodicity of collection of samples is suitably decided (Biradar, 2002). He further suggested that if the total number of fishing units is 10 or less, then data on all units should be collected, but if the fishing units exceed 10, a random sample of boats is selected. Sampling for fish length on a selected day should be such that it is representative of the catch landed on that day . Effort should be made to measure as many fish as possible preferably in thousands during a month and not in hundreds. Monthly length composition data so collected over different months in a year or over the years are processed for estimation of growth parameter (K), natural mortality (M), Fishing mortality (F), Yield per-recruit (Y/R) and Maximum Sustainable Yield (MSY).

A relevant sampling scheme should be made for other studies related to fish stock in terms of sexual maturity by size or age of fish, season and area, food and feeding (in multi-species context), and migration details.

It is absolutely important that sexing of fish should be done properly and sampling should be as dependable as possible. It is interesting to note that the wrong identification of sexes in the European eel, a fish whose biology has been most intensively investigated in Europe by British, Dutch, French, German and Italian scientists, led them to believe that eel changed sex. Also they believed that adverse conditions imposed maleness and favorable condition femaleness, since growth rate in the females was always higher than the males. Their basis of identification of sex was that the testis known as Syraski organ was morphologically lobulate whereas ovary was frilled. It was only when Sinha in 1965 examined the gonad histologically then it became clear that Syraski organ was not always diagnostic of maleness and was ,in fact, not always been the testis but small eels' ovary was also lobulate before becoming frilled ovary (Sinha and Jones, 1966).

It was also believed that the males and the females eels lived separately, males in the estuarine areas and females in the freshwater upstream. But when sampling of eel was done through electric fishing, almost all eels (smaller and larger) were fished out from the total stretch of certain streams in Wales, then it was found out that the males and females both lived near the estuary and also in the upstream contrary to the previous findings. The female eels migrate to sea from freshwater at the age of 16-18 years, whereas males at the age of 6-8 years. The previous investigators used long lines in freshwater where predominantly larger eels were caught. The samples of larger eels naturally were of older age groups and thus consisted of mostly females.

The smaller eels having lobulate organ were invariably found in the estuary where most of them were still having undifferentiated gonad but wrongly supposed (identified) to be the males. When the previous workers did not find the younger female eels with frilled gonad, they thought that since the female grow that fast, that it was difficult for them to find the younger female eels. This argument had no validity since the rate of growth may be faster, but that would not mean that they would skip any size group totally.

The fact that sex differentiation in eel occurs only at the age of 3-4 years, the samples containing smaller and younger eels with lobulate organs were wrongly identified as all males. Many of them were still undifferentiated, some with lobulate ovary and some with testis. Unfortunately, samples only for the age group of 4-8 years were never examined before to see the sex ratio and preponderance of a particular sex in a particular locality.

To explain the sex change in eel it was suggested that either the leptocephali on arrival to the estuary from Sargasso Sea selected their locality as per their future sex i.e., the males remained in estuary whereas the females went upstream, or they were distributed at random and favorable environment upstream imposed female sex whereas adverse conditions male sex.

It was also suggested that eel has unstable sex chromosome and is influenced either way because of environment. But the above findings with proper identification of sex histologically and with suitable method of collection of sample though electric fishing along with the selection of adequate sampling sites throughout the streams enabled to clarify the confusion of sex change in eel (Sinha and Jones, 1966). This highlights the importance of proper methodology of identification, collection of samples and the selection of sites for such investigations.

It is reiterated that sampling techniques and fishing gear are selective and sometimes, it becomes very difficult to quantify results.

Fish Stock Assessment Studies

An adequate sampling of the total stock or a management unit or all fisheries on the stock is needed to have stock assessment study by different methods such as :

(a) Surplus production models (Schaeffer and Fox) which require time series data on only catch and effort over the years to provide estimates of MSY.

(b) Analytical model of Beaverton and Holt (yield per recruit model) requires data on age or size composition of catch and provides estimates of optimum fishing effort and mesh size to obtain MSY.

(c) Virtual Population Analysis (VPA) or Cohort analysis also requires age or size composition data and provides estimates of fishing mortality (F) and population for different age or length groups.

(d) Thompson and Bell model requires input from VPA or Cohort analysis and average weight of age or length groups. This method is useful in predicting the effects of changes in the fishing effort on future yields. An important aspect of this model is that it allows for incorporation of value of the catch.

(e) Swept area method is based on research trawl survey catches per unit area and provides estimates of biomass and thus MSY.

The methods and models are either fishery independent e.g. area swept methods and acoustic methods or analytical fishery dependent models.

Length based Fish Stock Assessment (LFSA) package developed by FAO and the COMPLEAT ELEFAN package developed by ICLARM have been extensively used for fish stock assessment since late nineteen eighties. However, during early nineties a new package 'FAO-ICLARM Stock Assessment Tools (FiSAT) integrating positive aspects of both the packages has been developed and is commonly used in stock assessment studies.

Studies based on modeling approach also have opened up new avenues for the culture-based fisheries of small reservoirs (Sugunan and Sinha, 2000). Available models clearly indicate that production is a function of fishing mortality and stocking density. Certain standard variables on population parameters, such as the density-dependent growth, size dependent mortality and weight-length relationship when known; the optimum stocking density and the fishing mortality can be arrived at (Lorenzen, 1995)

Statistical Design for Experimentation

All the biological characteristics are known for their high variability and therefore, experimental response observations in any aspects of fish habit and habitat, however carefully planned, show high degree of variation. Hence, there is a need for application of statistics in designing, conducting and analyzing such experiments. Although, several experimental designs are available for the purpose, adoption of these designs in fisheries and aquaculture research has been very limited in most of the developing countries. Therefore, some preliminary basics that are useful and very relevant for experimental designs are indicated (Biradar, 2002).

The experimental error arises not only because of inherent variability in the experimental material but also due to failure of the researcher to standardize experimental techniques. This can be reduced adopting adequate replication and local control. Replication refers to repetition of treatment under investigation, while local control refers to grouping (blocking) of experimental objects into relatively homogeneous units. The number of replications or the units of objects or treatments should not be compromised unless advised by a statistician. In order to get unbiased estimates of treatment effects, treatments need to be allotted to experimental units by a random process called randomization.

Completely Randomized Design (CRD), Randomized Complete Block Design (RBD) and Latin Square Design (LSD) are the three basic designs for conducting statistically designed experiments. The CRD is mostly used for laboratory experiments, wherein each treatment is randomly applied to a few experimental units. RBD is suitable when there is only one source of variation that can be controlled by suitable blocking e.g. initial weight. In this design treatments are allocated randomly to experimental units within each block in such a way that each treatment appears once in each block. The number of blocks will be equal to number of replications in this design. LSD is used when there are two sources of variation say for example age and initial weight. In this design two restrictions are imposed by forming blocks in row and column-wise direction. Number of treatments equals number of replications in this design. Thus, each treatment appears once in each row and each column.

When the experiment involves a number of different levels of factors then factorial experiments are set up. For example, if the experiment consists of 3 stocking densities and 4 levels of water depth, then factorial experiment consisting 12 (=3x4) treatment combinations needs to be undertaken. The advantage of factorial experiments is that they provide information on interaction between various factors.

Non-biotechnological Studies

While all ecological and biological investigations should adopt relevant methodology to generate comparative and comprehensive data to arrive at meaningful conclusion, the studies on legal and social studies need standardization of methodologies and identification of quantifiable parameters. Public-good approach with societal benefits should be the underlying principle and the parameters that would help conduct investigation to achieve quantifiable results.

The methodology for economic analysis, particularly production studies involves estimating the resource use efficiency, comparing the costs and returns, analyzing the returns to scale, assess the frontier level of production, and evaluating the impact of technology by impact studies like (now and then or before and after) need at least a minimum of 30 sample to derive results of statistical significance and meaningful conclusions.

The important variables to be considered for the culture fisheries include the input and output parameters both in monetary as well as quantitative terms like yield, area, seed, feed, fertilizer, labor, disease control and prophylactics measures, depreciation, irrigation charges etc. Whereas, in capture fisheries important variables are the input and output parameters both in monetary as well as quantitative terms like yield, area, duration of fishing, type of craft, type of gear, distance from the bank or shore, fuel charges, frequency of fishing, experience of fishers, size of crew, quantity of discards, composition and size of the catch etc.

Analysis is normally done through correlation and regression studies, production function, linear programming or through advanced mathematic modeling, the expertise of which is rarely available in fisheries research institutes in developing countries. Therefore, at least simple but meaningful correlation and regression analysis must be attempted for such investigations.

At least 30 sample are to be collected to derive any result of statistical significance for marketing studies such as to compare the marketing efficiency, to analyze the market integration, to assess the price spread and marketing margin etc.. The important variables to be included in such studies are normally the prices at the different levels right from the producer, farm gate, wholesalers, retailer and consumer and also the marketing costs and margin in the different marketing channels.

The data in general are analyzed through the time series analysis or Ravallion models and depending on the objective of the studies other analytical methodologies are used such as Forecasting techniques, Demand and supply estimations, Seasonality Indices, Multiple Co-integration etc. (Shyam, Personal Communication).

5.3 Preparation of Research Project Proposal

The following outlines indicate the information required for a simple Proforma for writing any research project proposal

General Information

1. Institute name.
2. Name and address of the Research Unit/Center.
3. Title of the project.
4. Date of the start.
5. Likely date of completion.

Technical Information

6. Title of the Research Project.
7. Review of the work done in the past and justification (including the present state of the art wherever applicable and also an ex-ante analysis).
8. Goals and Objectives of the study.
   - Output.
   - Practical utility.
   - Environmental sustainability.
   - Social Equity.
   - Beneficiary.
9. Technical Program and details of the experimental design.
   - Research Methodology.
   - Research Monitoring and Evaluation Sheet.

Details of the Research Team

10. Name designation and experience of the project leader and of associates together with the time proposed to be devoted and work to be done.

Sl. No.	Name	Designation	Time to be spent (m/m)	Work to be done

11. Estimated man month
    - Scientific
    - Technical
    - Others

Details of the Requirements

12. Facilities available and additional requirement:
    - Land
    - Labor
    - Special equipment
    - Animal sheds
    - Fish ponds
    - Other items
    - Total estimated cost on new facilities
13. Inputs requirement
14. Inputs delivery schedule as per the Input Delivery Sheet

Details of the Budget

15. Detailed Budget
    - Salary of scientific staff
    - Salary of technical staff
    - Salary of other staff

- Cost of equipment, facility, and other inputs.
- Contingencies

Total estimated cost (approx.)

Approval by Concerned Officials

16. Signature of Project Leader and associates
17. Signature of Immediate Supervisor
18. Signature of Chief (Administration and Support services)
19. Signature of Director

Different organizations adopt different format, many of them are quite exhaustive. However, in general the project proposal should essentially indicate the background and justification of the proposal incorporating the review of the work done so far (as state of the art and also the ex-ante analysis) and its relevance in the context of national priorities. Followed by goals and the short, medium and long- term objectives, inputs requirements, output expected, institutional mechanism for conducting the studies, plan of studies, monitoring and evaluation mechanisms, input delivery schedule and the detailed budget.

Assessment of facilities and fund requirement

Proper analysis should be made for the assessment of facilities and fund requirement as shown in the Flowchart No 2.

Flowchart 2: Assessment of Facilities and Fund Requirement for Research Project Formulation (Sinha 2001)

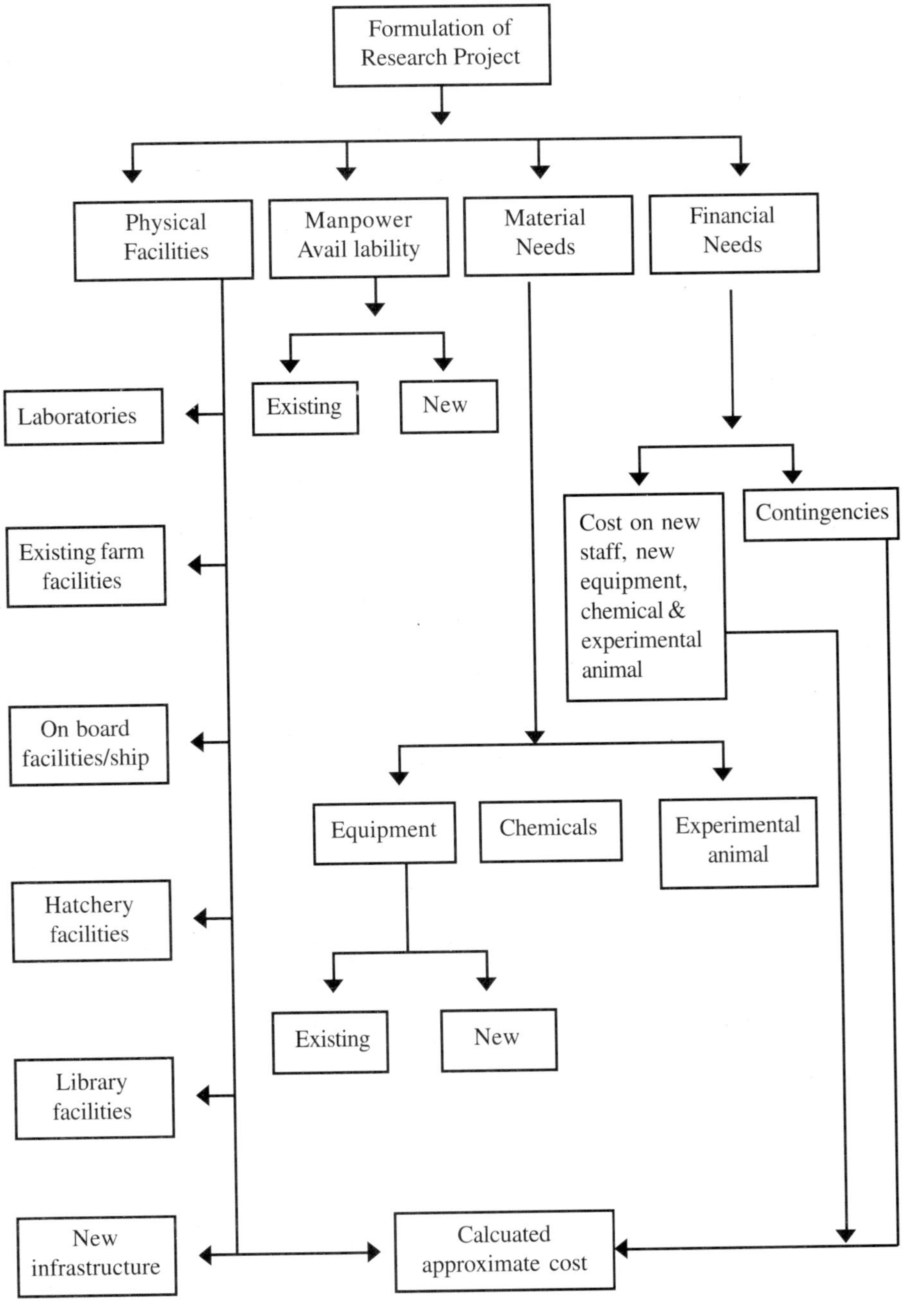

Preparation of Research Monitoring and Evaluation Sheet

It is important that every research proposal should show clearly the research activities and the time frame in which they should be completed. At the same time research activities need to have measurable parameters for timely monitoring. Therefore, the research proposal should include a Research Progress Monitoring and Evaluation Sheet (RMS) indicating as far as possible certain quantifiable criteria of the research activities such as specific parameters of investigation, sites and frequency of sampling, number of training program and trainees, number of workshop to be conducted, reports and research papers to be prepared during the research projects duration.

Simple examples of such sheets (Tables 4-7) to be included in the research proposals are shown here to provide a general guideline (Sinha, 2000).

Table 4. **Reproductive behavior and controlled breeding of certain endangered species: research activity, progress monitoring and evaluation sheet.**

Activity	Quantifiable criteria for monitoring	Time Frame (Year and Quarter)																				Results achieved	Specific constraints	Observation / Recommendation
		1				2				3				4				5						
		1	2	3	4	1	2	3	4	1	2	3	4	1	2	3	4	1	2	3	4			
1 Selection of the 10 most endangered fish	· Collection of endangered fish specimen from nature.	■	■																					
	· Raising of fish in controlled conditions			■	■	■	■	■	■	■	■	■	■	■	■									
	· Study growth, food and feeding habits in controlled condition			■	■	■	■	■	■	■														
2. Study Reproductive Biology of the endangered Fish species	· Investigate development of ovary and testes monthly and also the reproductive cycle	■	■	■	■																			

Activity	Quantifiable criteria for monitoring	Time Frame (Year and Quarter) 1				2				3				4				5				Results achieved	Specific constraints	Observation / Recommendation
		1	2	3	4	1	2	3	4	1	2	3	4	1	2	3	4	1	2	3	4			
	Investigate any sexual dimorphism, sexual differentiation and secondary sexual character.			■	■	■	■	■	■	■	■	■	■	■	■									
	· Investigate sexual and spawning behavior				■	■	■	■	■	■	■	■												
3. Maintenance of mother stock	· Undertake experiments on stocking density, nutritional requirement, and feeding with artificial feeding				■	■	■	■	■	■	■	■	■	■	■	■	■	■	■					
4. Production of seed under controlled condition in the hatchery	Undertake experiments on induced breeding of matured male and female					■	■	■	■	■	■	■	■	■	■	■	■	■	■					

Activity	Quantifiable criteria for monitoring	Time Frame (Year and Quarter) 1				2				3				4				5				Results achieved	Specific constraints	Observation / Recommendation
		1	2	3	4	1	2	3	4	1	2	3	4	1	2	3	4	1	2	3	4			
5. Raising of the seed	Undertake investigation on growth, food and feeding habits of fingerlings																							
8. Reporting	· Preparation of progress report (20)																							
	· Preparation of Annual report (5) · Preparation of research papers(4)																							
	· Preparation of terminal report (1)																							

Table 5. An economic study of production and marketing of fish from small scale Marine Fisheries: Research activity, progress monitoring and evaluation sheet.

Activity	Quantifiable criteria for monitoring	Time Frame (Year and Quarter)												Results achieved	Specific constraints	Observation / Recommendation
		1				2				3						
		1	2	3	4	1	2	3	4	1	2	3	4			
1. Survey and selection area	• Selection of specific areas for survey – primary, secondary and tertiary markets (atleast 3 each)															
2. Preparation of interview schedule and sample selection	• Preparation of questionnaire															
	• Development of sampling design															
3. Pre-testing and preparation of final schedule	• Pre-testing draft questionnaire															
	• Preparation of final schedule															
4. Data collection	• Collection of data for different gears in different locations and seasons															

5. Small-scale marine fishes	·Seasonal variation of marine catch
	·Species caught
	·Prepare list of major species
6. Production cost and returns	·Determination of cost and returns in marine fisheries
7. Marine fish marketing	·Select 5 major species for marketing survey
	·Select 5 depot, 10 processing factories and fish markets
	·Study on different locations and landing of marine fish and their marketing
8. Collection of information from Government officials and other secondary sources	· Meeting and collecting specific information from Government officials and other concerned people

Activity	Quantifiable criteria for monitoring	Time Frame (Year and Quarter)												Results achieved	Specific constraints	Observation / Recommendation
		1				2				3						
		1	2	3	4	1	2	3	4	1	2	3	4			
9. Checking of collected data	· Continuous checking and varification of collected data															
10. Tabulation and computerization	· Tabulation and data entry															
11. Analysis of data	· Tabulation and statistical analysis															
12. Reporting	· Preparation of progress report (12)															
	· Preparation of Annual report (3)															
	· Preparation of research papers (1)															
	· Preparation of Terminal report (1)															
	· Organization of workshop															

Table 6. Adaptive research on certain selected technologies of aquaculture: Research activity, progress monitoring and evaluation sheet.

Activity	Quantifiable criteria for monitoring	Time Frame (Year and Quarter)																				Results achieved	Specific constraints	Observation / Recommendation
		1				2				3				4				5						
		1	2	3	4	1	2	3	4	1	2	3	4	1	2	3	4	1	2	3	4			
1.Survey and identification of the farmers	· Preparation of proforma for survey	▬																						
	· Baseline survey of 200 farmers	▬	▬	▬	▬																			
	· Selection of 25 farmers for each technology each year (1-4)	▬				▬				▬				▬										
2. Selection of technology, technical program and preparation of ponds	· Outline step to be taken for management		▬				▬				▬				▬									
	· Pre-stocking management of 25 ponds			▬				▬				▬				▬								

Activity	Quantifiable criteria for monitoring	Time Frame (Year and Quarter)																				Results achieved	Specific constraints	Observation / Recommendation
		1				2				3				4				5						
		1	2	3	4	1	2	3	4	1	2	3	4	1	2	3	4	1	2	3	4			
3.Demonstration of technology	• Stocking of fish			■				■				■				■								
	• Study on growth of fish and other culture aspects			■	■	■	■	■	■	■	■	■	■	■	■	■	■	■	■					
4.Harvesting	• Maintainance of water quality			■	■	■	■	■	■	■	■	■	■	■	■	■	■	■	■	■				
	• Decision on harvesting				■				■				■				■							
	• Finilization of the plan for marketing					■				■				■				■						
	• Calculation of cost/benefits					■				■				■				■						

Activity	Sub-activity
5. Yearly workshop and finalisation of technical program and finalization of packaging of technology	· Organization of the workshop
	· Preparation of the report
	· Presentation of the proceedings and modification of the technical programme
6.Extension Program	· Organization of training programmes for 100 farmers each year
7.Reporting	· Preparation of progress report (18)
	· Preparation of Annual report (5)
	· Preparation of research papers (1)
	· Preparation of terminal report (1)

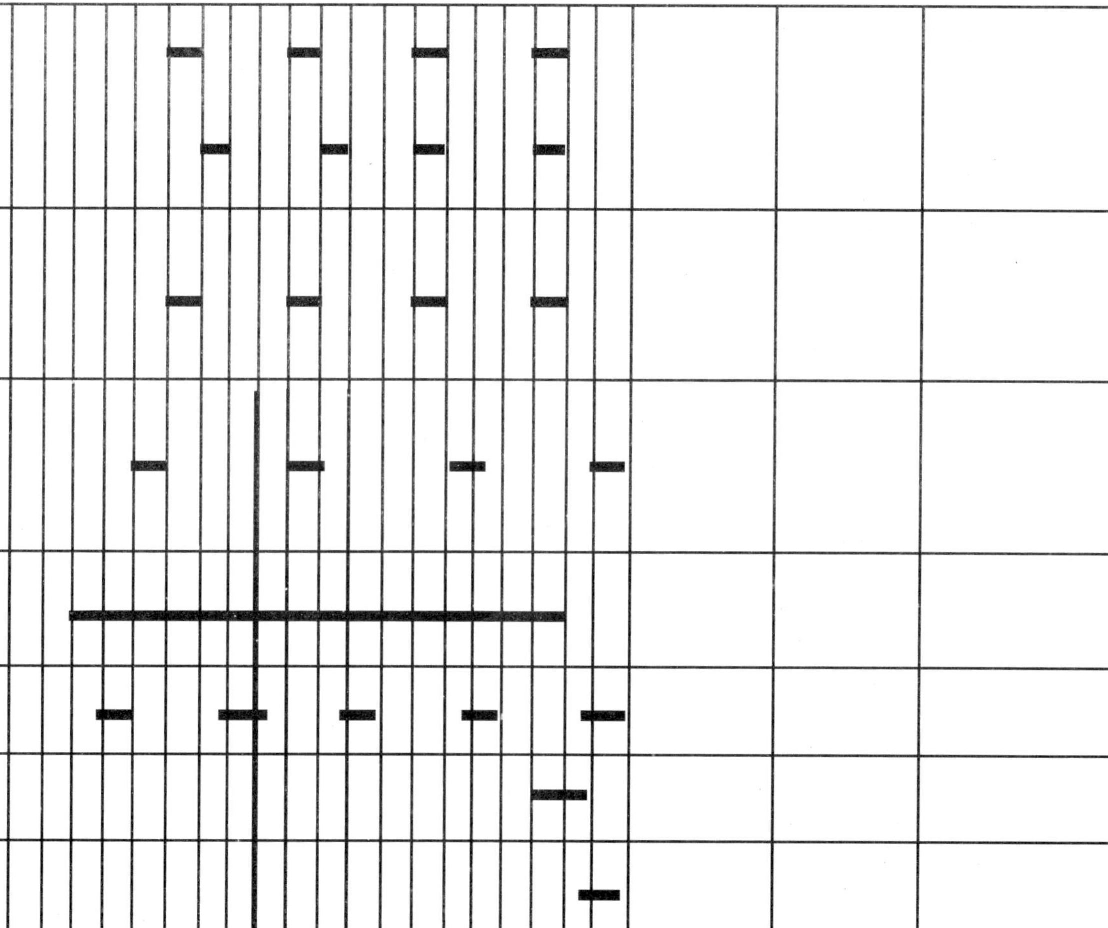

Table 7. Preliminary study on monitory of Hilsa fisheries in a particular river : Research activity, progress monitoring and evaluation sheet.

Activity	Quantifiable criteria for monitoring	Time Frame (Year and Quarter)																				Results achieved	Specific constraints	Observation/ Recommendation
		1				2				3				4				5						
		1	2	3	4	1	2	3	4	1	2	3	4	1	2	3	4	1	2	3	4			
1. Determination of age at maturity, frequency of spawning and sex differentiation	· Age determination of 30 fish each during spawning season in September/ October from six locations	■						■	■		■	■												
	· Histological study of gonad of 30 fish each from six locations during prestocking and post spawning season			■	■	■			■	■			■	■			■	■						

2. Monitoring of spawning and nursery grounds of Hilsa and study on physico chemical and biological parameters	• Major physico-chemical parameters pH, conductivity of water, salinity, freshwater supply and abundance of plankton quality and quantity to be assessed weekly before spawning and during spawning seasons
3. Deterimination of optimum mesh size for Hilsa fishing through gear selectivity study and study on key population parameters	• Monthly collection of data from 30 fish of each size groups from experimental fishing and from fishermen
4. Extension activities	• Training of farmers (200) in Jatka rehabilitation

Activity	Quantifiable criteria for monitoring	Time Frame (Year and Quarter)																				Results achieved	Specific constraints	Observation/ Recommendation
		1				2				3				4			5							
		1	2	3	4	1	2	3	4	1	2	3	4	1	2	3	4	1	2	3	4			
	· Workshop on Hilsa fisheries												■				■							
5. Reporting	· Preparation of the state of the art report on the Hilsa spawning grounds (1)									■														
	· Preparation of report on the Hilsa fisheries management and conservation (1)													■										
	· Progress report (20)	■	■	■	■	■	■	■	■	■	■	■	■	■	■	■	■	■	■	■	■			
	· Preparation of Annual report (5)				■				■				■				■				■			
	· Preparation of research papers (2)												■					■						
	· Preparation of terminal report (1)																				■			

Preparation of Input Delivery Sheet

It is equally important to monitor the status of input delivery. The input delivery should be properly planned in advance so that research activities should not suffer. A time schedule as shown in table-8 needs to be prepared for availability of each item of the inputs essential to conduct the research.

Table 8 Input Delivery Sheet for Research Project.

Year/Quarter		Staff recruitment		Infrastructure		Equipment		Financial			
		Progress	Constraints	Progress	Constraints	Progress	Constraints	Progress	Voucher Submitted amount/date	Any other specific constraints	Observation/re-commendation
	1										
	2										
1	3										
	4										
	1										
	2										
2	3										
	4										
	1										
	2										
3	3										
	4										
	1										
	2										
4	3										
	4										
	1										
	2										
5	3										
	4										

Steps for conducting research

To conclude, the following are the step-by-step exercise to prepare work plan and to conduct the research

1. Identify research problem according to national objectives and constraint analysis.
2. Literatures search and review of past work done including state of the art and ex-ante analyses.
3. Assess institutional capacities- institutional, financial and manpower.
4. Set research goals and objectives.
5. Organize seminar on research topic and discuss the past work and future proposed work.
6. Prepare detailed experimental design.
7. Prepare Research Monitoring and Evaluation Sheet.
8. Assess requirement of additional staff, equipment, chemicals and other experimental materials.
9. Prepare Time Schedule for Procurement (Inputs Delivery Sheet) of equipment, chemicals and other experimental materials.
10. Setting of experimental facilities.
11. Initiation of experiments.
12. Implement and manage the research study as planned.
13. Analyze short falls.
14. Finalize the results.
15. Complete documentation of research results.
16. Organization of seminar to present research results.
17. Prepare materials for extension and communication.
18. Publish research results.

5.4 Projects Prioritization

The resource allocation both in terms of fund and facilities for undertaking research is one of the most judicious tasks of the research managers. It is, therefore, important to undertake systematic and comprehensive methods of setting priorities among alternative research options and then decide on resource allocations. Based on cost and benefits, two types of assessment are done of the research studies.

(i) Ex-ante (before research begins) so as to select among alternative research studies/projects for allocating resources and particularly funding support.

(ii) Ex-post (after research) assessment for its impact.

Both *ex-ante* and *ex-post* assessments clearly indicate 'Now and Then' situations. While, *ex-ante* exercise normally assists in planning research and in estimating expected research pay off in future against the research cost, *ex-post* exercise indicates actual research pay off. These exercise help in guiding the research and technologies policies of the country.

There are five main methods for economic evaluation of agricultural research.

- Scoring
- Estimation of economic surplus
- Econometrics
- Mathematical programming
- Simulation models

The scoring method estimates the contributions of research project to its different objectives. In this method it is possible to calculate a single measure by giving different weights to different objectives to reflect their relative importance.

The economic surplus method estimates return to investment on research. Internal rate of return is estimated in term of benefits consumer and producer got with that of research cost. Falconi (1993) has further elaborated on this method. However, ex-ante analysis through this method would require information on expected value of production, expected yield increase, reduction of unit cost and/or the maintenance effects of research, probability of success of research, future market condition, adoption rate of the technology and appropriate discount rate for converting future benefits and cost into present value for comparing against the internal rate of return.

Similarly, econometrics, mathematical programming and simulation methods are complex and a host of data is needed on production functions and also for estimating supply, cost, and profit functions to estimate the contribution of research to total output. Marginal rate of return to research expenditure is used to prioritize the research studies.

These methods are also used to determine optimum size of the project for maximizing benefits.

Most of the developing countries lack expertise in fisheries resource economics, methodology, money and mechanism to have most reliable database. Similarly, fisheries researchers hardly take into account the existing situation in the field to help make ex-ante analysis and unless this exercise is done effectively, any reliable ex-post analysis is not possible. Therefore, research project formulation has to have ex-ante analysis as an integral part of it for any evaluation of the impact.

The Bangladesh Agriculture Research Council, Bangladesh undertakes prioritization of research projects as suggested by Collin and Kissi (1995) and Janssen and Kissi (1997). The process requires a fair assessment as far as possible of the expected benefits of each project. The assessment of benefits of each project is made through discussion and arriving at a consensus among the scientists, economists, transfer of technology experts and the research managers, and then the projects are assigned the ranking. During the project-ranking phase, candidate projects are compared for potential of benefits only and no consideration of manpower limitations is made (Morris, Personal communication). He further indicated the following steps for such prioritization of research projects:

Step 1. Establish tentative program areas for the next five years period.

Step 2a. Review and re-interpret recent past research relevant to each tentative program area: e.g., yields/rates versus potential yields/ rates at the national or agroecological zone (AEZ) or client group level, and unit costs of production.

Step 2b Identify constraints to production expansion as discussed in constraint analysis and rank the researchable problems to mitigate specific constraint and its impact on national goals.

Step 3. Define research projects within each program (on the basis of clusters of similar topics or related constraints), fix the duration of each project and specify the disciplines and scientist-years (SY) required to complete the project.

Step 4a. Determine expected annual rates of return per SY invested in the project using following equation, utilizing Economic Efficiency Index (Vx ri x si x zi x di)) and dividing by Scientists Year (SY). This gives the Research Priority Index (Ri)

$$Ri = (V \times ri \times si \times zi \times di) / SY$$

V is the current annual production of the commodity (local currency/year); ri is the annual rate of increase in production expected from the proposed research (i.e., contribution due to demand-pull are excluded); si is the expected probability of success of the project; zi is the proportion of V that is produced in the target clientele group of production area or AEZ, and di is the expected extent of adoption or diffusion within the target zi (after a lapse of 10 years).

The values of ri, si, zi and di are best estimated with the help of the experienced scientists and training/extension officers. The value of ri will be positive (proportionate yield change). The values of ri, si ,zi and di must be between 0 and 1. The above equation is divided by the scientist man months to have a Research Priority Index (R_i).

If desirable, weights for expected impact on employment, wealth creation, and food security can be inserted in the formula.

Step 4b. Rank each project according to expected returns to scientist-year with adjustment with success and adoption rates and policy factors dictated by goals of the national policies.

Step 5. Assess ranked projects against availability of scientists in the required disciplines, operational budgets, and facilities. Retain only those projects for which current personnel and facilities are adequate.

Step 6. The scientists responsible for those projects retained from Step 5 must develop the rationale and objectives; list the measurable indicators, and other requirements to record the Projects Research Management Information System (MIS) of the Institute.

As has been suggested by (Morris, Personal communication) the research projects priority setting at the Bangladesh Fisheries Research Institute, Bangladesh has been done through the simple scoring method. Five factors were mainly considered such as value of current annual production, expected yield increase annually because of proposed research, probability of research success, the proportion of total value of the production in the target clientele group or ecological zone and the expected rate of adoption i.e. diffusion rates (Sinha, 2001). Table 9 (➡) shows an example of such prioritization of research projects in Brackishwater fisheries sector.

Table 9. Brackishwater research project priority setting at Bangladesh Fisheries Research Institute (Sinha, 2001)

Name of the proposed projects	Expected rate of increase (ri)	Probability of success (si)	Area to be covered (zi)	Diffusion rate(di)	Annual Production of fish (million taka/yr) (Vi)	Scientific and other staff/y (SY)	Research Priority index ViXriXsiX-ziXdi/SY	Ranking
Studies of the soil and water quality to increase productivity in the Gher Fisheries.	0.20	0.50	0.30	0.20	37500	16.40	13.72	2
Development of nursery management and culture technique for commercially important shell and fin fish.	0.30	0.50	0.20	0.40	37500	15.12	29.75	1
Brood Stock management and Mass Seed production of P. *monodon* and fin fish	0.20	0.50	0.30	0.20	37500	18.88	11.92	4
Development of shell fish and finfish diet from locally available ingredients.	0.20	0.40	0.20	0.40	37500	17.90	13.41	3
Studies on the climatic aberration and its impact on shell fish and finfish reproductive biology	0.30	0.30	0.20	0.30	37500	18.63	10.87	6
Base line study on estuarine ecology and, biodiversity in different estuaries and rehabilitation of shrimp through ranching.	0.20	0.40	0.20	0.50	37500	36.550.3	8.21	7
Base line information and improvement of water quality to control fish disease epidemic	0.20	0.50	0.30	0.30	37500	30.50	11.07	5

The above prioritization has been done particularly because it is relatively easy to understand and apply on the basis of experience, expectations and the consensus among the scientists and managers.

It is far easier to make ex-ante and ex-post analyses when dealing with research on specific commodities with clear cut expected yields, value of production, incremental production and the proportion of the production in the particular agro-ecological zone, since they are the major parameters and outputs. But for fisheries research particularly in developing countries where focus of research needs to be on fish biodiversity conservation and maintenance, aquatic productivity management, soil and water quality maintenance and management and also on socio-economic aspects, above equation i.e. economic efficiency index does not seem to be adequate. It is essential that socio economic and environmental aspects should be considered of paramount importance in fisheries research. Therefore, it is recommended that setting priorities through scoring method should include placing weight also on these objectives.

To cite an example in Bangladesh situation the relative weight for socio economic gains (soi) in the scale of 1-10 has been tentatively put as: for employment for women 5, for food and nutrition of the poor as 2.5, and for employment for poor 2.5. Similarly, for environmental gains (ei) for water and soil water quality improvement 5, biodiversity conservation 2.5, rehabilitation and eco-restoration as 2.5 (as shown in the Flowchart 3). The sum of the allocated weights has been divided by 10, to derive a single measure to the respective objectives.

Flowchart 3 : Environmental and Socioeconomic Gains for Project Prioritization.

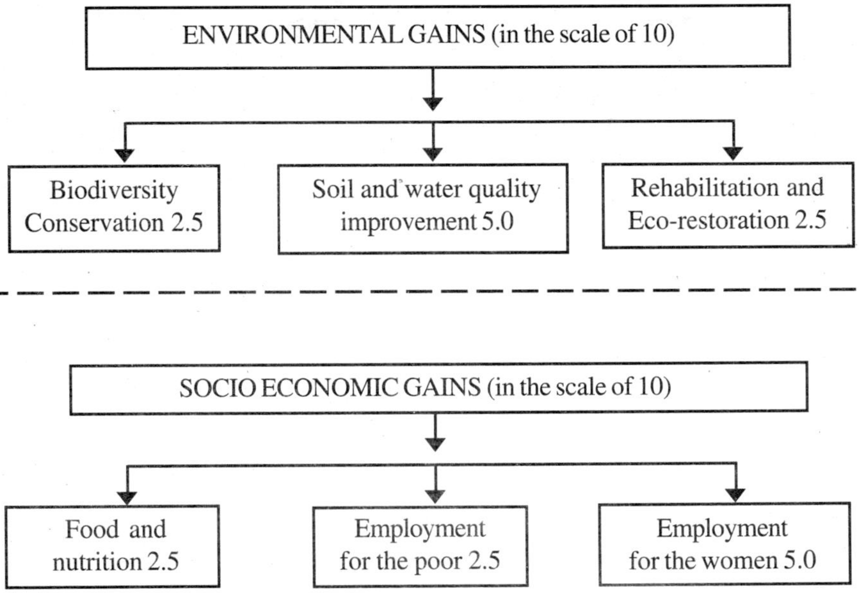

Thus, Rif - (Research Priority Index for Fisheries research) is to be calculated on the following formula:

$$Rif = V \times ri \times di \times zi \times si \times soi \times ei \div Sy$$

Where the

1) Economic gains are calculated as Economic efficiency index
2) Social gains are quantitative indicators. The (soi) is calculated as weighted index taking into account country specific national policy and priorities.
3) Environmental gains are compiled using arbitrary values based on scientific knowledge.

When the weights are assigned to these objectives and the Rif for the brackishwater fisheries research projects at BFRI is calculated, the ranking of the projects changes, which appears more valid and appropriate (Table 10. ➡) and fits in well with the required fisheries development policy and plan of the country (Mazid and Sinha, 1999 a and b).

Table 10. Comparison of ranking of brackishwater fisheries research projects (after inserting social and economic gains).

Name of the proposed projects	Ri=Vi/ Sy	Ranking of Projects with Ri	Social values (si)	Environmental values (vi)	Rif/Sy	**Final Ranking**
Studies of the soil and water quality to increase productivity in the Gher Fisheries.	13.7	**2**	0.5	0.5	3.42	**1**
Development of nursery management and culture technique for commercially important shell and fin fish.	29.7	**1**	0.4	0.2	2.38	**3**
Brood Stock management and Mass Seed production of P. *monodon*. and fin fhish.	11.9	**4**	0.4	0.6	2.86	**2**
Development of shell fish and finfish diet from locally available ingredients.	13.4	**3**	0.4	0.2	1.07	**7**
Studies on the climatic aberration (Global warming) and its impact on shell fish and finfish reproductive biology.	10.8	**6**	0.4	0.5	2.17	**5**
Base line information on estuarine ecology, biodiversity in different river estuaries and rehabilitation of shrimp through ranching.	8.2	**7**	0.3	0.5	1.23	**6**
Base line information and improvement of water quality to control fish disease epidemic.	11.0	**5**	0.4	0.5	2.20	**4**

Sometimes the gains for determining research contributions may be overlapping, such as socio-economic gains, which are also viewed as environmental gain. But scoring method has this inherent deficiency (Norton 1993).

Mruthynjaya (2001) discussed the importance of phases of the research-adoption-impact continuum if returns to investments in research are to be systematically quantified. Since the aim of research is to produce a new technology and improve or sustain farm productivity for improvement in the welfare of farmers and the society, attainment of the objective occurs in steps and in various phases of research such as basic, adaptive and applied. Achieving targets at each phase depends on research capability and research support. The probability of success in producing the final output depends on success of different phases.

The second aspect relates to changes in production, consumption and environment as technology is adopted and research results/products are utilized. Measuring impact (tangible and intangible) of new technologies – the third aspect of the continuum-on the welfare of farmers and society at the farm, regional and national level in general involves various considerations in addition to improving productivity and changing consumption pattern. Thus, he recommends using the following indicators.

- Relevance of new technologies to yield stability, food security, national honor and dignity.
- Effects of new technologies on health and environment.
- Sustainability of farming systems.
- Importance of low-input technologies among resource poor farmers.
- Gender roles in new technologies and their impact on family welfare.
- Effects on employment, input use, poverty alleviation, trade (import and export).
- Distributional effect on society.
- The scientific knowledge generated besides impact of technology generated.

He cited an example of ex-ante economic evaluation of hybrid rice research in India. The analysis estimated net present value of benefits (NPB) and internal rate of returns (IRR) arising from the expected adoption of hybrid rice in irrigated areas in India.. The *ex-ante* economic evaluation compared the costs and expected benefits of hybrid rice research program. It used both the scientific and economic parameters.

The economic surplus model was used to estimate the total benefits and their distribution among the producers and consumers. The model was used in a closed economy framework with linear supply and demand functions and parallel research induced supply shift. The result of the analysis indicated that the NPB is about Rs. 30 billion showing an internal rate of return of 66%. The producers and consumers would share research benefits almost equally.

There are indirect benefits on account of additional employment generation particularly for women in hybrid seed production. It is estimated that about 237 man-days of incremental employment is generated by per ha. of hybrid rice seed production. Since 93% of the incremental work in seed production is done by rural women, it would promote the entrepreneurial and gender equity. Besides this, seed conditioning and sales would also generate additional employment opportunities.

A multi –objective scoring model was developed for research priority setting in the semi-arid tropics, the details of which are given by ICRISAT (1992) and Joshi and Bantilan (1997). Ahmed *et al.* (1999) also indicated certain aspects of scope and methodologies of assessment of the impact of aquatic resources research.

Recently a framework for setting priorities for fisheries research has been indicated by Saxena *et al.,* (2002), which is of great interest. However, it will depend on the expertise available with the fisheries institutions in the developing countries. Generalized fisheries research priorities in South Asia as finalized during a meeting of Asia Pacific Association of Agricultural Research Institutions (APAARI) and quoted by Saxena *et al.,* (2002), needs a serious review in view of discussions made in Chapter 4.

6. Research Monitoring, Management and Evaluation

The mechanism of monitoring and managing the research project should enable the research manager and administrator to keep the track of progress of research and circumvent any constraint in its progress. It should assist the researchers to achieve the objectives of their research in time and within the approved budget. It should also provide the opportunity for mid- course correction if needed and help to expedite the input delivery. Horton *et al.* (1993) have given valuable suggestions on monitoring and evaluation of agriculture research.

6.1 Frame Work of Monitoring and Management of Research

Even though, many research activities are very difficult to have measurable parameters for timely monitoring but certain quantifiable criteria need to be developed to monitor the progress. It is important as has been suggested in the last chapter that every researcher should prepare a Research Monitoring and Evaluation Sheet (RMS) and include in the research proposal. The RMS will guide the researcher about the progress of research.

It is important to reiterate here again that constraints of fund, facilities and manpower will continue to exist in varying degree in different institutions particularly in developing countries and therefore it becomes the important duty of every scientist and supervisor to make the best use of the existing resources. It is equally important to monitor the status of input delivery. The input delivery should be properly planned in advance so that research activities should not suffer. At times non-performance of the scientists is directly related to non- availability of inputs in time. However, the following simple measures will improve the situation considerably.

- Realistic advance planning by the Principal investigator (PI) for input delivery in consultation with the officials of scientific support service.
- Submission of indent by PI for equipment and chemicals with detailed specification to the Store and Procurement Section much earlier than the initiation of research studies keeping in mind the time of annual approval of the budget and availability of fund.
- Immediate installation of new equipment and outmost care of their maintenance, in no case equipment should be kept unused and kept for longer period in the store having no air-conditioning.
- Timely submission of voucher for fund recoupment, it is important that advances should be settled in time and care should be taken to avoid any unnecessary delay. Financial discipline is very important for research management and hence to be strictly observed.

The simple practices as stated above, if meticulously adopted at the PI level, will go a long way in successful implementation of the research proposals. It is also very important that commitment of the inputs by the concerned officials should be judiciously made and delivered. In fact, only those projects should be finally approved by the Director /Chief of the Institute where manpower and facilities exist and funding is almost ensured. This is a pre-requisite for successful management of the research (Sinha, 2001).

A general framework for management and monitoring of research is suggested here, which consists of external monitoring as well as internal and self-monitoring. It is expected that the Governing Body (GB) of the institution indicates over all research priority areas in conformity with national development plan and policy and ensures that the research projects reflect those. It is important that the GB procures the required fund and provide facilities for the research. The duties of the GB have been dealt in the Chapter 3. However, a technical group should advise the GB about the relevance of research studies in terms of national priorities and the technical aspects involved in a particular study.

Technical Advisory Body

It is always desirable that the institutions should have a Technical Advisory Body (TAB) consisting of eminent scientists, and different stakeholders to assess the relevance and technical soundness of the research projects. TAB should be guided strictly by the opinion of the subject matter specialists. A roaster of such specialists should be made in all institutions, from where they can be picked up for assessing the technical aspect of the project.

The subject matter specialist should rate the research projects in ascending scale of 1-10 on the following aspects and make recommendation on relative technical soundness.

Table 11. Shows the proforma of technical rating of the research project

Different Aspects of the Project	Scale 1-10
1. Relevance of the project to the national development plan	
2. Previous work done in the proposed area	
3. Scientific merit and quality of the research project proposal	
4. Appropriateness of the research methodology and technical program	
5. Appropriateness of the Research Monitoring and Evaluation Sheet	
6. Likely achievements of the objectives within the time framework and budget indicated	
7. Expected economic, environmental and social benefits	
8. Availability of the necessary research facilities and infrastructure	
9. Team work and collaboration	
10. Justification of facilities and funds proposed a. Manpower support (For Research and Research related activities) b. Renovation or creation of infrastructure c. Budget	
11. Recommendation	**Total**

On the basis of the above recommendation the Technical Advisory Body needs to examine the research proposal on the following aspects:

- Examine the remarks of subject matter specialist.
- Relevance of the research against the existing constraints and emerging opportunities in the development of fisheries sector.
- Examine duplication of such research works in other institutions, if any.
- Suggest and indicate if item of work is essential and fruitful.
- Make specific recommendations of additional manpower, infrastructure facilities and budget considering the usefulness of research studies and expected economic and other benefits.
- Recommendation and technical approval of research proposal.

  Subsequent review of the progress of project proposals approved earlier.

Internal Monitoring

This step involves final approval of the project and budget after the technical clearance of the Technical Advisory Body. The Chief Executive of the institute should approve the project with necessary budget and staff, and evaluate the progress and the quality of research work.

The internal monitoring by the Director/ Chief of the Institution involves the following :

- Approval of the project with necessary budget and monitoring of expenditure.
- Provide continuous research support and monitor the input delivery.
- Monitor and evaluate the quality of research progress and output.
- Assess and approve further need of research.
- Decide whether research procedure needs to be modified or study needs mid course correction.
- Take a decision on the utility of research results.
- Approve packaging of research results into technology.
- Publish annual report and undertake other related reporting matter.
- Approve the manuscript of research papers.
- Take a decision on initiating action for patenting fisheries product & biodiversity.

The internal monitoring is a continuous process and it is the lifeline for successful and cost effective research studies. In order to facilitate this, Research Monitoring and Evaluation Sheet should be meticulously updated and made available in time to concerned senior official (Sinha, 2000).

Monitoring by Head of the Division

The Head of the Division should make a realistic assessment of facilities, manpower and fund while recommending the project proposals to the chief executive of the institute. Certain constraints are likely to exist and thus it warrants highly judicious decision in selecting the project proposals. But once they are recommended and finally approved, the Head of the Division has to ensure that the scientists get the facilities, manpower and fund in time, besides making sure of his/her own involvement as an active researcher and as a leader in his/her discipline. It needs to be ensured that no researcher should make any excuse of such constrained situation for non-delivery of the project outputs. Thus, the following functions are to be carried out at the head of the division level.

- Close supervision of research as per proposal.
- Ensuring timely provision of inputs under constrained situation and prevent scientists making excuse of non-performance because of such constraints.
- Maintain and monitor quality of research.
- Update the Director on the progress of research both technical as well financial.
- Packaging of research results in time.
- Prepare annual report.
- Review scientific paper and recommend for publication.

PI level / leader of the team

- Undertake leadership role in conducting the research.
- Conduct research and motivate co-workers to perform well under constrained situation.
- Complete the research component as indicated in the Research Monitoring and Evaluation Sheet.

- Assist in getting the inputs in time and also to solve the constraints in research progress.
- Prepare progress report both technical and financial and submit in time.
- Prepare manuscript of scientific paper.

Self Monitoring by the Scientist

- Undertake research as per the project proposal.
- Assist PI in sorting out the problem of input delivery.

The scientist working in the laboratory or in the field has the sole responsibility to conduct the research as committed in the research project proposal. The RMS will be the best guide to monitor the progress of research.

6.2 Research Management Information System

The International Services for National Agriculture Research (ISNAR) has developed a Research Management Information System (RMIS) that it makes available to National Agriculture Research System (NARS). The RMIS, named INFORM-R, is based on a widely used rational database management system: MS access 97. By adopting this MIS, the time required to create the structure of a new RMIS is saved. Furthermore, the structure of INFORM-R is based on ISNAR's experience with various NARS in different countries. It has been made user-friendly in the sense that major computer expertise is not required to install, maintain or use the system. As ISNAR improves INFORM-R, new versions will be available to the agricultural research institutions (Morris, Personal communication).

RMIS provides essentially information on the research structure composed of Program-Project-Activity, the details of the initial plans and their progressive completion during the life of the project, utilization of the fund and facilities and manpower for various processes. This information system provides direct access of the information to monitor research activities, existing constraints in execution of the research, status of funding and input supply and at the end the final research outputs of the project. A periodic assessment of actual progress versus those indicated in the RMS is entered in the RMIS to show research mangers how efficiently project scientists are using resources including their time, knowledge and skills as well as institute resources to achieve the objectives agreed on at the outset of the project (Morris, Personal Communication).

6.3 Evaluation of Research Project

Monitoring and evaluation are the two sides of the same coin, complementing each other activities in several ways. It is important to understand that monitoring is a continuous process of gathering, analyzing and reporting the research studies, for the purpose of managing and identifying problems at an early stage. It also helps in the financial management and in monitoring of supply of the inputs in time.

The RMS becomes the best tool for monitoring of any research project since it sets quantitative performance targets such as the steps to be taken to achieve the research objectives, the quantum of work to be undertaken, relevant parameters to be investigated, number and frequency of observations to be made etc.

Progress should be discussed regularly in suitable forum. Periodic evaluation of the research progress is not only important in achieving the targets but finding out the reasons in case if they are getting difficult to achieve. Then it becomes easier for the research manager to undertake mid course correction to improve the progress.

Besides the satisfactory progress of the project, publication of research papers, organization of workshops, number of technology evolved and training of the beneficiaries need to be considered for scientific evaluation. However, the following aspects also reflect on the performance of the researchers:

- Quality of Benchmark survey or ex-ante and state of art analyses;
- Submission of progress reports in time;
- Quality of research study reflected in research publications;
- Maintenance and management of the laboratories and farms;
- Group work and coordination with other agency;
- Quality of the seminar ;
- Appreciation of the needs of the farmers, fisherman and women;

It is important to reiterate that as far as possible a proper benchmark study or ex-ante and the state-of art analyses should be made before the research project is started so that a realistic assessment of the impact of research can be made through ex-post analysis.

7. Research Results Packaging and Testing

7.1 Research Results Packaging, and Evolution of Technology

Science and technology are closely related, in fact, science is the basis of a technology. A technology is evolved when scientific research results are integrated for the end user. But every research study does not necessarily result in giving rise to a technology. In fact, a number of studies is required to develop scientific practices of aquaculture, packaging of which, results in the formation of a technology. For example, technology for fish seed production involves packaging of the practices of brood stock management, knowledge of the spawning habit, technique of induced breeding, method of incubation of the eggs, hatchery management and maintenance, larval development, its food and feeding, nursery raising, method of harvesting and transportation and also information on economics of input and output. Similarly, technology for rice-fish culture involves packaging of the practices and information on different aspects such as understanding of the ecosystem, preparation of rice field ecosystem, stocking of fry for fingerling production, its harvesting management, culture of fish in rice field and its harvest and its economic evaluation.

Similarly, a number of scientific studies is required to evolve guideline formulation for open-water fisheries management or ecosystem restoration or fisheries rehabilitation . The following detailed examples clearly indicate the quantum of scientific work needed for certain aquaculture technologies package (Sinha, 2000a).

1. Technology Package for Asiatic Carp Seed Production.

Research Studies	Research Result
• Studies on brood stock management, raising of brood stock, their appropriate stocking density, understanding the breeding cycle, identification and selection of the mature mother-stock, nutritional requirement of brood fish and feeding and their health management.	Brood Stock Management
• Observation on the secondary sexual character, spawning habit, breeding behavior, courtship, spawning techniques specific to fish.	Spawning Habit
• Induced breeding through hypophysation, breeding/spawning, gonadal hydration.	Controlled Breeding
• Stripping of males, fertilization of eggs, water hardening, care of eggs, identification of good and bad eggs, oxygenation of eggs, hatchery management.	Fertilization & Incubation of Egg and Hatchery Management
• Larval feed requirements, feeds formulation and preparation.	Larval Feed
• Nursery preparation, eradication of weed, insects and mollusks and fertilization.	Nursery Preparation
• Stocking, stocking density, duration of raising, feeding, health and disease management.	Stocking Management
• Harvest, conditioning/acclimatization, oxygen packing. Method and mode of transportation.	Transportation
• Economics of input and output.	Economics of carp seed production

2. Technology Package for Rice-fish Culture.

Research Studies	Research Result
• Studies on commonly occurring fish fauna in rice fields, desirable species, and their food and feeding habit. Commonly occurring insects in the field, their role and control.	Fish in Rice Field Ecosystem
• Preparation of rice fields for raising of the seed or undertaking fish culture. Suitability of trenches, enclosures and fish hide out, water control structure, and appropriate water level during and after weeding. Method of safe application of pesticides and fertilizers. Measures to reduce the effect of pesticides.	Preparation of Rice Field
• Stocking density, and period of fry raising for fingerling production.	Fingerlings Production
• Methods of harvesting, packing and transportation. Profitability of seed raising.	Harvesting Management
• Screening of suitable fish species for culture, their stocking proportion, stocking density, duration of culture.	Stocking of Fish for Culture
• Method of harvesting, Fabrication of suitable traps and gears. Profitability and economics of rice-fish culture.	Harvest Management

3. Technology Package for Carp Cultivation in Pond.

Research Study	Research Result
• Method of pond preparation, requirement of fertilizers depending on pond soil and water nutrient status, role and application of lime, method of composting, recycling of nutrients in the pond bottom, oxygen balance.	Pre stocking Management

• Size, density, and proportion of fingerlings of different species, and relative advantage of polyculture.	Stocking Management
• Nutritional requirement and effects of environmental factors, like temperature, oxygen, over crowding, different types of artificial feeds, method of feeding, broadcasting, platform feeding, feeding device, feeding schedule.	Feeding Management
• Management and maintenance of water and soil quality, control of algal and bacterial bloom.	Maintenance of water and soil quality
• Common fish disease prevention and cure, method of treatment.	Fish Health Management
• Harvesting, different method of harvesting, time of harvesting, duration of harvesting, complete single harvesting or multiple harvesting.	Harvesting Management
• Transportation and marketing. Economics and profitability of fish culture.	Transportation & Economics of carp culture

4. Technology Package for Integrated Fish-Livestock Farming.

Research Study	Research Result
• Relationship between different sub production systems. Different methods of integration. Designs and construction of integrated fish farm. Selection of plants which should be cultivated in the farm and on pond embankment, their farming techniques.	Method of Integration

• Manurial value of pond mud, and of bird dropping. Plant and animal waste treatment, method of decomposition and composting.	Fertilization Management
• Preparation of the pond, selection of the fish species, stocking density and proportion.	Preparation of the Pond
• Nutrient status of soil and water, the amount of organic manure requirement and assessment of the number of birds to be raised in the pen.	Bird Raising
• Duck/chicken rearing, egg production and marketing. Vegetable/crop production and marketing. Fish harvesting and marketing. Profitability of integrated fish farming.	Marketing and economics of fish-livestock farming

5. Technology Package for Cage Culture.

Research Study	Research Result
• Studies on different types of cages, floating cage, rigid and flexible cage, self supported and raft supported. Construction of cage, framework, support and lifting ropes, walls, bottom and roof, selection of materials for framework. Mooring and anchoring, Selection of site and installation of cage, Construction of water breaking structures.	Construction of Cages of Different Type
• Selection of fish species, stocking rates. Nutritional requirement, different type of feed, feeding of fish, feeding frequency and methods. Maintenance and cleaning of nets.	Stocking Management

• Prevention and control of disease and predators.	Disease and Predator Control
• Selection of size at the time of harvest, harvesting. Profitability of cage culture.	Harvest and Profitability

6. Technology Package for Shrimp Culture.

Research Study	Research Result
• Studies on design, construction and maintenance of farm on the basis of soil quality, land elevation, salinity, temperature, and tidal amplitude.	Selection of Site and Design of Pond
• Studies on preparation of pond, sun drying, ploughing and leveling of pond bottom, liming and application of organic and inorganic fertilizers and eradication of predators.	Pre stocking Management
• Studies on the stocking size/age, stocking density of PL.	Stocking Management
• Studies on the pond management, aeration and change of water, stock management and manipulation, production and maintenance of natural food, plankton production.	Post Stocking Management
• Studies on food and feeding, different types of feed, feeding rate and frequency.	Feeds and Feeding Management
• Studies on water quality management, control of plankton bloom, water temperature, p^H, control of disease.	Water Quality and Health Management
• Methods of harvesting and post harvest handling of shrimps. Profitability of shrimp farming.	Harvest Management and economics

Similarly, for open water fisheries management a number of studies is required to formulate policy guidelines for its conservation, management and development. For example, guidelines for hilsa fisheries management will require studies on different aspects of biology of hilsa particularly on its growth, feeding and feeding relationship within fish communities, breeding behavior, migration, spawning, larval survival, population structure at a given time and space, hilsa fishing, craft and gear used, degree of juvenile destruction, adult harvest and handling, and marketing etc along with hilsa fisherman attitude and skill about its fisheries. Only on detailed studies it could be possible to develop hilsa fisheries management technology.

7.2 Testing and Transfer of Technology

On the basis of the on-station research studies the technologies are packaged and tested in the field. The tested technologies are to be evaluated by the nodal department or agencies before it goes the field. Different countries have different set up and institutions dealing with the approval of technology to be transferred on the national scale. It is always desirable that a pilot scale testing of the technology is done in the field with the active participation of stakeholders. And then finally the technology should be disseminated to the end users. The Indian Council of Agricultural Research initiated in India a highly successful approach of generation and dissemination of technology in agricultural, animal sciences and fisheries. The scientific studies conducted in its research institutions invariably formed the basis of packaging of technology which were tested, refined, and adopted on the national scale through a number of All- India Coordinated Research Projects with active collaboration with the State Agricultural Universities and the State Governments. It initiated few All Coordinated Research Projects on Fisheries in 1971. The most outstanding performance of the Composite Fish Culture (Polyculture of Major Carps both Indian and Chinese) and Seed Production Project catalyzed revolutionizing carp culture in India and to some extent in the subcontinent through training of senior aquaculturists from Asian Member- countries of Network of Aquaculture Centers In Asia, then a regional project of FAO/UNDP (Sinha1999).

It is important to understand that mere research study or result or technology in itself cannot bring about any development of the sector unless it is supported by the government policy and, its effective implementation. Institutional support for inputs supply, credits, and marketing infrastructure need to be in place along with the proper mechanism for transfer of technology. Similarly, any scientific guideline for fisheries management needs its strict implementation and enforcement by the government and appreciation, participation and compliance by the stackholders, otherwise it is highly frustrating for the scientists to watch helplessly their research endeavor going waste. The flow chart on R& D explains the components required for fisheries development.

Flowchart 4 : On Research & Development in Fisheries Sector (Sinha, 2001)

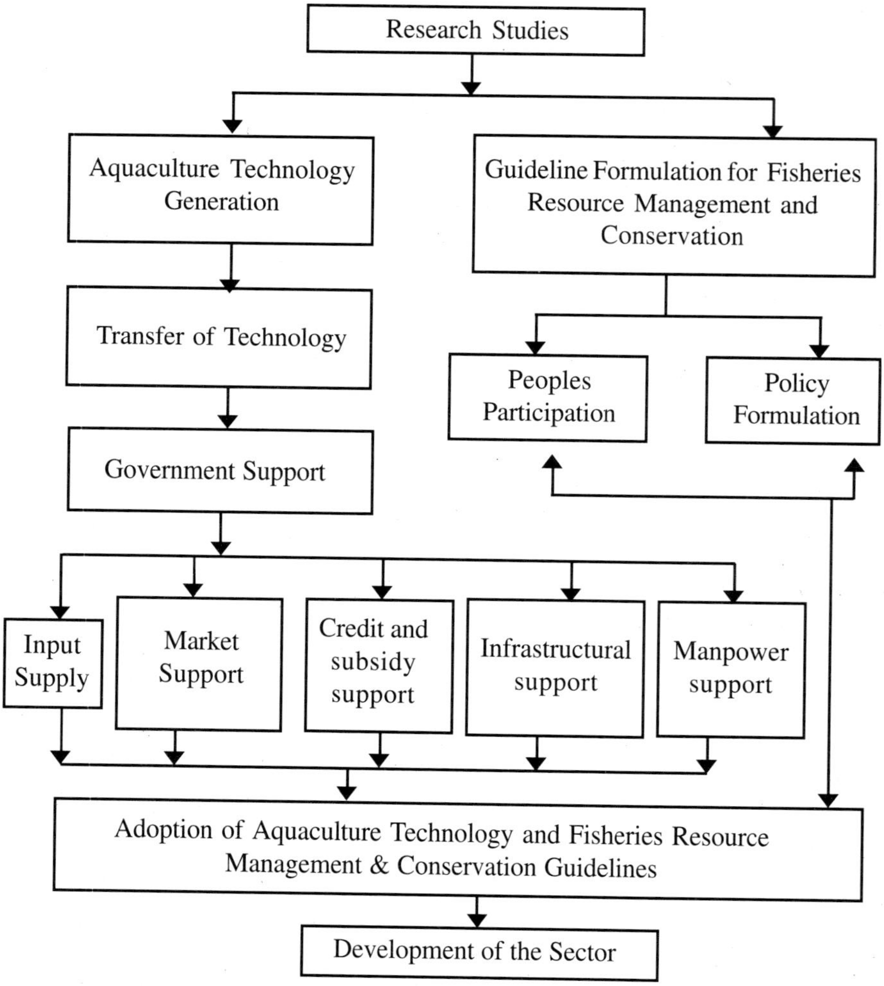

8. Capacity Enhancement in Education, Extension and Training

8.1 Capacity Enhancement in Fisheries Education

With the emergence of new frontiers of Fisheries Science and Technology and a higher expectation of increased production from the sector, an in depth examination is essential to reassess the quality of fisheries education and training, to meet the present requirement and priorities of the country concerned. In general, it appears that adequate human resource needs to be developed both in terms of quality as well as quantity at the primary, secondary and tertiary levels for fisheries development.

It is seen in many cases that institutions have been set up with massive infrastructure, without creating proper manpower. But even where manpower exists there is a serious mismatch in terms of quantity and quality, creating difficulty in achieving the desired mandate of the institution. Slow growth of fisheries in many developing countries, in spite of high resource potential, availability of technology and government support, appears to be due to the lack of properly trained and qualified manpower, who could imbibe and practice modern fisheries technologies and scientific management to help accelerate development. Fisheries industries find fisheries graduates and trained personnel ill equipped to match their needs (Chua, 1986). The present fisheries education and training give greater stress on descriptive aspects of fish biology, ecology and biological production process, based on the classical concept of zoology/biology. This results in putting limitation on the fisheries graduates to meet the modern demand of the industry. The sector needs more of market and resource management with environment sustainability and social equity.

The ICLARM (1986) and the Asian Fisheries Society (1988) generated valuable information on different aspect of fisheries education and training in Asian countries. Further, Shetty (2002) has given an overview of fisheries education in the Asia – Pacific Region.

Some of the Asian countries have established modern training facilities. But many developing countries have yet to establish adequate fisheries education facilities. The curricula in most of the institutions are either inadequate or suffer from lack of practical component. In some countries there has been a full-scale importation of course curricula of developed countries, thus ignoring the vital aspects of local requirements. Most of these have completely neglected the component of management sciences, especially of business, marketing and personnel management.

There is a serious lack of suitable text books, manuals and well-equipped laboratories. Most institutions need up-gradation. There is also a need for revision and modernization of the course content from time to time, up-gradation of staff qualifications, besides creating suitable infrastructure for on-the-job field training. However, now it is encouraging to note that many developing countries have initiated and strengthened their fisheries education in selected disciplines at the tertiary level, besides offering integrated courses for professionals at secondary level and organizing effective training for fishermen and fish farmers at the primary level (Sinha *et al* 1992, Shetty, 2002).

8.2 Future Thrust in Capacity Enhancement

8.2.1 Fisheries Education

Fishermen and fish farmers normally lack any formal fisheries education and therefore it is important that primary education at school level is imparted to create awareness of the fisheries potential in terms of resource conservation and utilization, farming system, processing and product development. They need to be trained through vocational courses in scientific innovations and techniques for better utilization and conservation of aquatic resources. They should be motivated to know how to supervise and manage them so that they remain as a healthy self-generating replenishable resource of their livelihood.

The undergraduate fisheries education is required to produce the much needed technically competent extension personnel to undertake transfer of technology, technicians to assist in research and development, instructors to teach in schools and technical institutions besides field managers and administrators to execute fisheries development projects.

Similarly, the postgraduate education at the tertiary level is needed to provide research scientists, higher executives and planners with larger disciplinal base to undertake relevant research and development works under government and non-government organizations and industries, in addition to providing teachers for universities, institutions of advanced learning and centers of excellence.

In general, the course curricula should be carefully framed focusing resource management with stress on the study of ecology in relation to impact of several extraneous as well as fishery activities on the aquatic resources. Similarly, the productivity enhancement should give more stress on genetic trait improvement, growth, feed conversion, disease control etc. Courses on market and business management should give lessons on advances made in managing business enterprises and marketing of perishable commodities, horticultural and livestock processing and product development etc. Study of cooperatives and its management is to be included in the course to create awareness and drive home the potential advantage of cooperation for management of fisheries of the common property resource and also for collective production and marketing.

It is generally seen that the courses on fisheries do not have very strong thrust on economics, socio-economic and legal aspects, partly because these disciplines have received very little research support to develop into a full-fledged fisheries discipline. A thrust in teaching these aspects through case studies would equip the students with a better appreciation of the importance of the subjects.

The role of communication and extension for effective transfer of technologies cannot be over emphasized. This is one of the most vulnerable areas, needing outmost attention. Suitable courses need to be incorporated on fisheries extension, journalism and communication.

Similar courses for secondary and primary levels need to be developed suitably encompassing wide spectrum of relevant topics considering the target group's knowledge and comprehension.

In India, when the Central Institute of Fisheries Education, Mumbai became the first fisheries university in 1989 in the Indian sub-continent the course syllabus of M.Sc and Ph.D were developed considering the biological, ecological and marketing aspects. The following tables provide an outline of the courses, which will be useful in designing the course contents of fisheries educational institution or department in many developing countries.

Course contents of M.Sc. (Fisheries Management) (Four Semester course), at CIFE, Deemed University, 1989

Course No.	Title of the Course	Credit Hours Theory + Practical
FM 501	TAXONOMY, BIOLOGY AND ECOLOGY OF FIN FISH	1 + 1
FM 502	TAXONOMY, BIOLOGY AND ECOLOGY OF SHELL FISH	1 + 1
FM 503	MARINE FISHERY RESOURCES	1 + 0
FM 504	INLAND FISHERY RESOURCES	1 + 0
FM 505	LIMNOLOGY AND MANAGEMENT OF INLAND WATERS	1 + 1
FM 506	INLAND AND COASTAL AQUACULTURE	1 + 1
FM 507	AQUACULTURE DEVELOPMENT AND MANAGEMENT	1 + 0
FM 508	FISH NUTRITION, FEEDS AND FEEDING	1 + 1
FM 509	CULTURE-CUM-CAPTURE FISHERIES	1 + 1
FM 510	FISH DISEASES AND THEIR CONTROL	1 + 1
FM 511	OCEANOGRAPHY	1 + 1

FM 512	METEOROLOGY AND HYDROGRAPHY	1 + 1
FM 513	MARINE BIOLOGY	1 + 1
FM 514	AQUATIC POLLUTION, CAUSES, EFFECTS AND CONTROL	1 + 0
FM 515	POPULATION DYNAMICS, STOCK ASSESSMENT AND MANAGEMENT	1 + 1
FM 516	FISH HARVESTING TECHNOLOGY	1 + 1
FM 517	FISHING CRAFTS	1 + 1
FM 518	FISH PROCESSING TECHNOLOGY AND PRODUCTS DEVELOPMENT	1 + 1
FM 519	FISH GENETICS	1 + 1
FM 520	FISHERY ECONOMICS AND MARKETING MANAGEMENT	1 + 1
FM 521	FISHERY LEGISLATION AND LAW OF THE SEA	1 + 0
FM 522	BASIC MATHEMATICS	1 + 1
FM 523	BASIC STATISTICS	1 + 1
FM 524	COMPUTER PROGRAMMING	1 + 1
FM 525	PLAN AND PROJECT FORMULATION, IMPLEMENTATION AND EVALUATION	1 + 0
FM 526	MANPOWER REQUIREMENT AND HUMAN RESOURCE DEVELOPMENT	1 + 0
FM527/I FAM20	FISHERIES EXTENSION	1 +1

FM528/ IFAM 524	PRINCIPLES OF BUSINESS MANAGEMENT AS APPLIED TO FISHERIES	1 + 0
FM529/ IFAM 529	MANAGEMENT ACCOUNTING AS APPLIED TO FISHERIES	1 + 1

MINIMUM CREDIT REQUIREMENTS FOR M.Sc. DEGREE

COURSE WORK	—	32 CREDITS
RESEARCH WORK	—	16 CREDITS
SEMINARS	—	2 CREDITS

TOTAL 50 CREDITS

Course Contents of M.Sc. (Inland Fisheries Administration and Management) at CIFE, Deemed University, 1989 (Four semester course).

Course No.	**Title of the Course**	**Credit Hours Theory + Practical**
IFAM 501	TAXONOMY, BIOLOGY AND ECOLOGY OF FIN FISH	1 + 1
IFAM 502	TAXONOMY, BIOLOGY AND ECOLOGY OF SHELL FISH	1 + 1
IFAM 503	PATHOLOGY AND PARASITOLOGY OF FIN FISH AND SHELL FISH	1 + 1
IFAM 504	FRESHWATER AQUACULTURE	1 + 1
IFAM 505	BRACKISHWATER AQUACULTURE	1 + 1

IFAM 506	TEMPERATE REGION AQUACULTURE	1 + 0
IFAM 507	AQUARIUM AND SPORT FISHERIES MANAGEMENT	1 + 0
IFAM 508	FISH BASED INTEGRATED CULTURE SYSTEMS	1 + 0
IFAM 509	INLAND AQUATIC ECOSYSTEMS AND THEIR MANAGEMENT	1 + 0
IFAM 510	AQUATIC POLLUTION AND ITS MANAGEMENT	1 + 1
IFAM 511	FISHERIES STATISTICS AND COMPUTER PROGRAMMES	2 + 1
IFAM 512	REMOTE SENSING IN INLAND FISHERIES	1 + 0
IFAM 513	REVERING AND LACUSTRINE FISHERIES	1 + 0
IFAM 514	ESTUARINE AND LAGOON FISHERIES	1 + 0
IFAM 515	INLAND FISHERY CRAFT AND GEAR	1 + 1
IFAM 516	AQUACULTURE ENGINEERING	1 + 1
IFAM 517	FISH FARM MACHINERY MAINTENANCE	1 + 1
IFAM 518	ECONOMIC ASPECTS OF INLAND FISHERIES	1 + 0
IFAM 519	FIN FISH AND SHELL FISH HANDLING PROCESSING AND PRESERVATION	1 + 1
IFAM 520	FISHERIES EXTENSION	1 + 1

IFAM 521	INLAND FISH MARKETING MANAGEMENT	1 + 1
IFAM 522	INLAND FISHERIES CO-OPERATIVES	1 + 0
IFAM 523	RISK AND INSURANCE IN INLAND FISHERIES	1 + 0
IFAM 524	PRINCIPLES OF BUSINSES MANAGEMENT AS APPLIED TO FISHERIES	1 + 0
IFAM 525	FISHERIES ADMINISTRATION IN STATES AND CENTRE	1 + 0
IFAM 526	FISHERIES POLICIES PLANNING AND PROJECT FORMULATION	1 + 0
IFAM 527	PROJECT IMPLEMENTATION MONITORING AND EVALUATION	1 +0
IFAM 528	MANPOWER REQUIREMENT AND HUMAN RESOURCE DEVELOPMENT	1 + 0
IFAM 529	MANAGEMENT ACCOUNTING AS APPLIED TO FISHERIES	1 + 0
IFAM 530/ FM 519	FISH GENETICS	1 + 1

MINIMUM CREDIT REQUIREMENTS FOR M.Sc. DEGREE

COURSE WORK	—-	32 CREDITS
RESEARCH WORK	—-	16 CREDITS
SEMINARS	—-	2 CREDITS

TOTAL 50 CREDITS

Course Contents Syllabus for Ph.D. programme (Fish & Fisheries Science) (6 Semester course), at CIFE, Deemed University, 1989.

Course No.	Title of the Course	Credit Hours Theory + Practical
I.F. 602/ F.F. 601	TAXONOMY, BIOLOGY AND ECOLOGY OF INLAND FIN FISH AND SHELL FISH	2 + 1
M.F. 603/ F.F. 604	TAXONOMY, BIOLOGY AND ECOLOGY OF INLAND MARINE FIN FISHES AND SHELL FISHES	2 + 1
I.F. 605/ F.F. 603	SOIL, WATER AND PRODUCTIVITY MANAGEMENT	2 + 1
I.F. 606/ F.F. 604	ENVIRONMENT MONITORING, POLLUTION AND FISH DISEASE	2 + 1
M.F. 605/ F.F.605	MARINE AND ESTUARINE POLLUTION	1 + 1
M.F. 606/ F.F.606	MICROBIOLOGY AND FISH DISEASES	2 + 1
I.F. 607/ F.F.607	REPRODUCTIVE PHYSIOLOGY, GENETICS AND BREEDING	3 + 1
I.F. 608 F.F. 608	FISH DIGESTIVE PHYSIOLOGY, NUTRITION AND FEED	3 + 1
I.F. 609/ F.F.609	BIOTECHNOLOGY	2 + 1
I.F. 612/ F.F.610	INLAND AQUATIC ECOSYSTEMS AND THEIR MANAGEMENT	1 + 0
I.F. 613/ F.F. 611	INLAND OPENWATER FISHERIES	1 + 0

M.F. 604/ F.F.612	MARINE AND ESTUARINE ECOLOGY	1 + 1
F.F.	RECENT ADVANCES IN AQUACULTURE SYSTEM	1 + 1
M.F. 604/ F.F.612	MARINE AND ESTUARINE ECOLOGY	1 + 1
F.F. 613	RECENT ADVANCES IN AQUACULTURE SYSTEM	1 + 1
M.F. 610/ F.F. 614	MARICULTURE	1 + 1
M.F. 613/ F.F. 615	MARINE FISHERY MACHINERY AND INSTRUMENTATION	2 + 1
I.F. 614/ F.F. 616	FIN FISH AND SHELL FISH HANDLING, PROCESSING AND PRESERVATION	1 + 1
I.F. 615/ F.F. 617	INLAND FISHERIES CO-OPERATIVES	1 + 0
I.F. 616/ F.F. 618	FISHERIES STATISTICS, COMPUTER PROGRAMMING AND REMOTE SENSING	1 + 1
Fy.Econ. 613/	FISHERIES FINANCE AND CO-OPERATION	1 + 1
Fy.Econ. 605/ F.F. 620	FISHERIES PRODUCTION AND RESOURCES ECONOMICS	2 + 0
Fy.Econ. 606/ F.F. 621	INTERNATIONAL LAW OF THE SEA AND COMMISSIONS	1 + 0

Fy.Econ. 607/ F.F. 622	FISHERIES LEGISLATION AND REGULATION IN INDIA	1 + 1
Fy.Econ. 612/ F.F. 623	FISH MARKETING AND PRICE ANALYSIS	1 + 1
Fy.Econ. 614/ F.F. 624	SOCIO-ECONOMIC ASPECTS OF FISHERMEN	1 + 0
Fy.Ext. 601/ F.F. 625	EXTENSION ORGANISATION SYSTEMS	2 +0
Fy.Econ. 606/ F.F. 626	EXTENSION ADMINISTRATION SYSTEMS	2 + 1
Fy.Econ. 608/ F.F. 627	RECENT TRENDS IN EXTENSION EDUCATION	1 + 0
Fy.Econ. 612/ F.F. 628	TRAINING AND PERSONNEL MANAGEMENT FOR FISHERIES DEVELOPMENT	2 + 1

MINIMUM CREDIT REQUIREMENTS FOR Ph.D. DEGREE

CORUSE WORK			CREDITS
- MAJOR FIELD	:	FISH & FISHERIES SCIENCE	18
- MINOR FIELD(2)	:	FISHERIES EXTENSION EDUCATION }	12
- (ANY TWO)	:	FISHERIES ECONOMICS }	(6+6)
	:	FISH PROCESSING }	
CREDIT SEMINAR			02
(Not on the topic of thesis)		THESIS RESEARCH	45
			TOTAL 77

The above courses have recently been further modified (Mukherjee - Penonal Communication)

8.2.2 *Fisheries Training*

It is essential to plan and organize suitable training programs for fishermen and fish farmers, who normally lack modern skills for enhancement of aquatic productivity, management and maintenance of resources and also for marketing of their produce. Similarly, suitable training is required for extension personnel, technicians and instructors. They need to know the principle and procedure involved in technologies concerned with productivity enhancement, maintenance of resources, and management of market, so that they can assimilate those and suitably use their knowledge with necessary modifications according to the needs of the clienteles. Senior managers and administrators who are involved in the planning, and executing fisheries development plans and projects should undertake study tours in and out of the country to get better acquainted about the potentiality and constraints of the sector and the methods of circumventing those.

At the tertiary level, research scientists need training on methodological research to improve their skill in undertaking research with uniform methodology and assays for comparable results with others. They should also specialize in relevant fields to acquire deep understanding of the subject and contribute to the advancement of the knowledge.

There is a serious need of organizing suitable training programs for effective transfer of relevant technologies from institution to field, from one region to another and also from one country to another. The training requirements and the topics will vary from country to country and will be commodity and site specific, yet an attempt has been made to indicate the broad outline of course contents of some of the important training programs for farmers, extension workers and researchers. These will provide a valuable guideline to organize effective training programs by the different institutions (Appendix I).

8.2.3 *Fisheries Extension Education*

An extension service is an educational process to educate, train, and motivate the end-users to accept innovations or improved technologies or certain recommended package of practices for raising their productivity. It is done by various means and methods: (i) individual contacts, (ii) group contacts, and (iii) mass-contacts. In order to make these methods effective, several communication aids and devices are used: photographs, posters, films, slides, filmstrips, video-cassettes on one hand and the publications like hand bills , leaflets, magazines, bulletins etc. on the other. A combination of these methods and aids has always been found most effective. The demonstrations of both "method" and "result" are the leading mechanisms for rapid spread of technologies among the clienteles.

Every country has some sort of extension system or the other for transfer of technology (ToT), The system are variously named in different countries. Cooperative Extension System in the USA; Farm Advisory Services in the European countries; Training & Visit System (T&V) in several developing countries promoted by the World Bank, and Community Development (C.D.), National Extension Services; T&V and First-line Extension System in India.

The First-line Extension System is an unique extension model promoted by the Indian Council of Agricultural Research (ICAR) — as an integral part of the National Agriculture Research System (NARS) in India. The underlying principle is to bring as soon as possible the new and latest technologies coming out of the research efforts to the notice of the farmers and the fishermen as also to the extension workers and agencies of the governments. The general extension systems of the Ministry of Agriculture, Government of India and the State Governments are mainly involved in the diffusion and spread of the technologies in the country at large.

The Indian Council of Agricultural Research has several projects under the first-line extension system in India: National Demonstrations, Lab to Land Programme, Operational Research Projects and the Farm Science Centers (Krishi Vigyan Kendras — KVKs) establishment in different districts. Now, all these projects, for the sake of efficiency as well as economy, were integrated into Farm Science Centers (FSCs). The FSCs/KVKs have the following mandates : (i) organize vocational, entrepreneurial and in-service training; (ii) conduct national demonstrations on the farmers fields; (iii) promote lab to land programme among the poorer sections of the farmers — marginal and small ; (iv) conduct operational/on-farm extension activities including exhibitions, farmers' fairs, animals shows and various farm competitions.

The KVK/FSC has successfully served the Indian agricultural sector including fisheries. There are now over 300 such centers in the country and more are being established in the remaining 200 districts. This model with over three decades of experience is highly relevant to other developing countries.

8.3 Fisheries Manpower Planning

A critical examination of the fisheries disciplines as set out in Section 3.2.1. does indicate the acute shortage of fisheries specialists. It is important that concerned governments should look at it and develop the human resource development programs to carry out needed research, education and training. The manpower requirement should be assessed considering the number of research, education and extension agencies, their present manpower, their personnel policy, present and future disciplinal requirement, national expectation from fisheries and government policy etc. A brainstorming session on "Manpower requirements and HRD in fisheries sector' held at CIFE, Mumbai, in 2000", discussed, at length, various aspects of man power requirement in fisheries sector in India (Ayyappan – personal communication).

The quantification of adequate manpower to meet the requirement at the primary, secondary and tertiary levels is difficult as indicated before, since it depends on several factors. Yet, a tentative assessment of trained manpower requirement in India based on the concept of unit operations for vessels and boats for open water fisheries, quantity of seed production in hatchery, area/ production for aquaculture, and processing plants of varying capacity has been attempted by Sinha *et al* (1992). Kohli (1998) did another exercise on such requirement. These exercises are by no means valid for every situation but will help others in estimating to certain extent the manpower requirements at different levels and accordingly in human resource planning for fisheries.

8.4 Linkage of Institutions with Center of Excellence

In order to develop major centers of fisheries education and research at the tertiary level in the developing countries, it is essential to select the potential institutions in those countries and establish their functional linkage with institutes, and universities in other countries having centers of excellence on fisheries disciplines.

The national institutions should upgrade their faculty and scientists through training them at those advanced centers of excellence/universities/institutions mainly in fish productivity enhancement research such as in fish ecology, fish genetics, fish reproduction, fish physiology, fish nutrition, fish microbiology, fish disease, fish bio technology and fish biochemistry etc.

Concurrently, the national institutions should concentrate on their capacity enhancement on management research, both of the resources and of the market. This will help the institution in forming a multidisciplinary team on return of those who get trained abroad in fish productivity enhancement research. It is important that the researchers should specialize in the ecological and market management as far as possible under national conditions rather than in an alien environment abroad with entirely different ecological and socio economic conditions.

To facilitate establishment of linkage with centers of excellence a list of institutions is given below in different subjects; primarily for fish productivity enhancement study. However, the list is not a complete directory of such centers and also does not reflect any shortcoming of those centers, which are not mentioned.

List of Centers of Excellence/ Institute /University for Possible Linkage

1. Fish Behavior

The University of Guelph, Guelph, Ontario, Canada

College of Agriculture and Natural Resources, Michigan State University, East Lansing, MI, USA

2. Fish Biochemistry & Biotechnology

Agriculture and Natural Resources Department, Storrs, CT, USA

University of Hampshire, New Hampshire, USA

Dept. of Fisheries & Applied Aquaculture, Auburn University, Alabama, USA

Birmingham College of Food & Domestic Arts, Birmingham, UK

Loughborough University of Technology, Loughborough, UK

University of Hull, Hull, UK

Loughry College of Agriculture & Food Technology, Loughry, Northern Ireland

Central Institute of Fisheries Education, Mumbai, India

Fisheries Research Institute, Szarvas, Hungary

Academia Sinica, Wuhan, China

Chinese Academy of Science, University of Hanjzhou, Hanjzhou, China

Asian Institute of Technology, Bangkok, Thailand

Aquatic Animal Health Research Institute, Bangkok, Thailand

University Pertanian, Kuala Lumpur, Malaysia

University of Uppsala, Sweden,

Ludwid Maximillians, Universitat Munich, Munich, Germany

University of Bergen, Bergen, Norway

Nagasaki University, Nagasaki, Japan

Tokyo University of Fisheries, Tokyo, Japan

Shimonoseki University of Fisheries, Japan

Zoology Department, University of Singapore, Singapore

3. Fish Ecology.

Dept. of Fisheries and Applied Aquaculture, Auburn University, Alabama, USA

University of New Hampshire, New Hampshire, USA

Columbia University, New York, USA

Cornell University, Ithaca, New York, USA

School of Aquatic and Fishery Science, University of Washington, Washington, USA

Vermont University, Burlington, Vermont, USA

Kentucky State University, Frankfort, Kentucky, USA

Purdue University, West Lafayette, Indiana, USA,

Institute of Fishery Resource Ecology, Florida State University, Gainesville, Florida, USA

College of the Environment and Life Science, University of Rhodes Island, Kingston, USA

Virginia Polytechnic Institute and State University, Blackburghs, Virginia, USA

Southern Illinois University at Carbondale, IL, USA

Wildlife and Fisheries Management Institute, Texas Tech. Institute, Lubbock, TX, USA

Texas A&M University, Houston, Texas, USA,

University of Maryland, Frostburg, MD, USA

College of Agricultural and Natural Resources, Michigan State University, East Lansing, MI, USA,

Kansas State University, Manhattan, KS, USA,

School of Natural Resources College of Agriculture and Live Science, University of Arizona, Arizona, USA

Biology Department, University of California, Davis, CA.USA

Biology Department , San Diego State University, San Diego, CA,USA

College of Natural Resources, Colorado State University, Fort Collins, CO, USA

Coastal Fisheries Institute, Louisiana State University, Baton Rouge, Louisiana, USA

The University of British Columbia, Vancouver, British Columbia , Canada

The University of Guelph, Guelph, Ontario, Canada

Memorial University of Newfoundland St. John's, Newfoundland, Canada

The University of Waterloo, Waterloo, Ontario, Canada

Regional University of Cariri, Crato, Brazil

Federal University of Parana, Centro, Curitibia, Brazil

National University of Loja, Loja, Equador

University of Southampton, Southampton, UK

University of Wales, Bangor, UK

University of Notingham, Nottinghamshirs, UK

University College, London, UK

Department of Ecology and Environment, Liverpool University, Liverpool, UK

University College, Galway, Ireland

Laboratory for Ecology and Aquaculture, Naamstraat 59,3000, Leuven, Belgium

Department of Aquaculture, Bodo University, Bodo,Norway

Waggeningen Agricultural University, Netherlands

Universite de Perpignan, Perpignan, France

Brackishwater Aquaculture & Development Centre, Jepara, Indonesia

Ocean Research Institute, Tokyo University, Tokyo, Japan

Kagoshima University, Japan

University of Philippines in the Visayas, Philippines

4. Fish Endocrinology

Southern Illinois University at Carbondale,IL. USA

University of Sheffield, Sheffield, UK

University of Waterloo, Waterloo, Canada

Dept. of Fisheries & Applied Aquaculture, Auburn, Alabama, USA

Nagasaki University, Nagasaki, Japan

University of Delhi, Delhi, India

5. Fish Disease

Dept. of Fisheries & Applied Aquaculture, Auburn University, Alabama, US

Kansas State University, Manhattan, KS,USA

Fish Aquatic Animal Health Research Institute, Bangkok, Thailand

National Inland Fisheries Institute, Bangkhen, Bangkok, Thailand

University Pertanian, Kuala Lumpur, Malaysia

Central Institute of Freshwater Aquaculture, Bhubaneswar, India

University of Plymouth, Plymouth, U.K.

Institute of Aquaculture, University of Stirling, Scotland, UK

6. Fish Genetics

Kansas State University, Manhattan, KS, USA

College of Agricultural and Natural Resources, Michigan State University, East Lansing, MI, USA

School of Fisheries and Ocean Sciences, University of Alaska, Fairbanks, Alaska

Southern Illinois University at Carbondale, IL, USA

Zoology Department, Oklahoma State University, Stillwater, OK, USA

Institute of Aquaculture, University of Stirling, Scotland, UK

Dept. of Genetics, University of Liverpool, UK

University of Reading, Reading, UK

Institute of Marine Research, Bergan, Norway

Institute of Aquculture Research, Akvaforsk, 1432 Aas, Norway

CSIRO, Marine Research Laboratory, Cleveland, Australia

University of Santiniketan, Santiniketan, West Bengal, India

Kamraj University, Madurai, India

Bureau of Fish Genetics Resources, Lucknow, India

Asian Institute of Technology, Bangkok, Thailand

7. Fish Microbiology

Fisheries Research Institute, Szarvas, Hungary

Central Institute of Fisheries Technology, Cochin, India

University of New Hampshire, New Hampshire, USA

Zoology Department, Oklahoma State University, Stillwater, OK, USA

Birmingham College of Food & Domestic Arts, Birmingham, UK

Loughborough University of Technology, Loughborough, UK

Loughry College of Agriculture & Food Technology, Loughry, Northern Ireland

8. Fish Nutrition

Fisheries Research Institute, Szarvas, Hungary

Dept. of Fisheries & Applied Aquaculture Auburn, University, Alabama, USA

College of Agricultural and Natural Resources, Michigan State University, East Lansing, MI,USA

Central Institute of Freshwater Aquaculture, Bhubaneshwar, India

Central Institute of Fisheries Education, Mumbai, India

Kiyoso University, Fukuoaka, Japan

Fish Nutrition Lab., Unite Mixte, Inra-1F Remer, 64310 Saint Peer-Sur-Nivette, France

University of Plymouth, Plymouth, U.K.

9. Fish Physiology

Kansas State University, Manhattan, KS,USA

Dept. of Fisheries & Applied Aquaculture, Auburn University, Alabama, USA

College of Agricultural and Natural Resources, Michigan State University, East Lansing, MI,USA

10. Fish Reproduction

Institute of Aquaculture, University of Stirling, Scotland, UK

Dept. Genetics, University of Liverpool, UK

Kamraj University, Madurai, India

Institute of Marine Research, Bergan, Norway

Institute of Aquculture Research, Akvaforsk, 1432 Aas, Norway

CSIRO, Marine Research Laboratory, Cleveland, Australia

11. Fish Taxonomy

University of Tokyo, Tokyo, Japan

University of Newcastle, Newcastle-upon-Tyne, UK

Southern Illinois University at Carbondale, IL, USA

8.5 Linkage of Institutions for Training

The national institutes should be fully involved in training their nationals at primary and secondary levels in appropriate technologies for progressive development of fisheries. These institutions should also concentrate on developing teaching aids and technology manuals for the training of teachers, extension workers, technicians, executives, fish farmers and fishermen.

A close collaboration among the developing nations in skill development at secondary level in different aquaculture systems will speed up the spread of aquaculture. Asian countries with a long aquaculture tradition are full of experience of successes and failures and thus they offer opportunities to Africa and Latin America to take advantages of the successes and learn lessons from the failures.

Culture of carp, tilapia and catfish has been highly successful in Bangladesh, China, India , Indonesia, the Philippines and Thailand. While catfish culture in Thailand did very well particularly of *Pangasius sutchi*, which occurs in Cambodia, Laos, Thailand and Vietnam, *Chana stritus* is showing promising trend in India and Vietnam. Paddy cum fish culture is undertaken very successfully in Japan, China, Indonesia, Thailand and in Africa in Madagascar. Integrated fish farming with crop-livestock and fish has been highly successful in China and is getting popular in many other countries like Bangladesh, China, India, Laos. Indonesia, Vietnam, Thailand, Nepal, and Pakistan. Similarly, Cameroon, Cote d'Ivoire, Kenya, Madagascar, Nigeria and Zambia have shown encouraging results.

The culture based capture fisheries in flood plains, and culture of fish in cages and pens have been very successful in China, the Philippines and Bangladesh. However, in Africa, traditional culture based fisheries such as in 'fish holes' in Cameroon and Tongo and 'Whedos" or 'Houedo" in Benin, 'Howash' in Egypt, 'Brush-park (Acadja)' in Ghana, Benin and Nigeria need improvement with scientific innovations.

It is desirable that the relevant institutions in Africa should develop linkage with advanced centers in Asian countries for training their secondary levels clienteles.

The African Institutions such as the Institute des Savanes (IDESSA), Bouake, Cote d'Ivoirs, the African Regional Aquaculture Centre(ARAC), Port Harcourt, Nigeria, the Institute of Animal and Veterinary Research(IRZV), Foumban, Cameroon, the Faculty of Agriculture, University of Buttare, Rwanda, the Department of Fisheries, Chilanga, Lusaka, Zambia and others should establish linkages between the Central Institute of Freshwater Aquaculture, Bhubaneswar in India and also between the Bangladesh Fisheries Research Institute, Mymensingh , Bangladesh for carp, catfish and tilapia culture and seed production.

Similarly, the following African Institutions may have a linkage with the Asia-Pacific Regional Reseach and Training Centre in Integrated Fish Farming, Wuxi, China, the National Inland Fisheries Institute and Asian institute of Technology, Bangkok, Thailand for integrated fish farming and also with Directorate of Fisheries, Jakarta, Indonesia for paddy cum fish culture.

Harvest of an Indian major carp Catla (Courtesy of Nath and Bhowmick)

Harvest of River Catfish, Pangasius in India (Courtesy of Nath and Bhowmick)

Harvest of Catfish, Clarius, in Bangladesh (Courtesy of Mazid and Hussain)

Harvest of Red Tilapia, (Courtesy of Mazid and Hussain)

Harvest of Freshwater Prawn from a pond in BFRI (Courtesy of Mazid and Hussain)

MERH, Ambatolampy Fish Farm, Madagascar.

Department of National Parks and Wildlife Management and University of Zimbabwe, Zimbabwe.

Lake Kariba Research Station, Zimbabwe.

Institute of Animal and Veterinary Research(IRZV), Foumban, Cameroon.

National Fish Culture Centre, Landjia, Central African Republic.

Nigerian Institute for Freshwater Fisheries Research, New Bussa, Nigeria.

It is desirable that the following African institution should establish linkage for culture based capture fisheries with the Central Inland Fisheries Research Institute, Barrackpore, India and with the Bangladesh Fisheries Research Institute, Bangladesh. Highly successful experience of Bangladesh in flood plain fisheries rehabilitation is most relevant.

Institute des Savanes (IDESSA), Bouake, Cote d'Ivoirs

Institute of Aquatic Biology, Accra, Ghana

Sagana Fish Culture Research Station, Kenya

Mcllwaine Research Station, Harare, Zimbabwe

Tanzanian Fisheries Research Institute, Dar-es-Salaam, Tanzania

Shrimp culture has been wide spread in Asian countries particularly in Thailand, China, Taiwan (China), India, the Philippines, Indonesia, Vietnam and Bangladesh and Republic of Korea. Latin American countries particularly Ecuador and Chile have done exceedingly well. However African countries have yet to make a real mark. It is important that intensified efforts should be made to expand coastal farming in Africa for which linkage is essential with the Southeast Asian Fisheries Development Center, Iloilo, Philippines, the Central Marine Fisheries Research Institute, Cochin, India and the Central Institute of Brackishwater Aquaculture, Chennai, India.

Mariculture is also highly developed in Asia particularly in China, Japan, Republic of Korea, Indonesia, Taiwan (China), Malaysia, the Philippines and Thailand. While Taiwan has developed its oyster culture, Malaysia concentrated on cockles and oysters. Philippines produce mussels and oysters. Thailand produces mussels, cockles and oysters But in Africa except farming of loup (*Dicentrachus labrax*) and daurade (*Sparus auratus*) in Tunisia, abalone (*Haliotis midae*) in South Africa, oyster in Morocco, South Africa, Seychelles and Tunisia, mussel in South Africa and Tunisia and seaweed in Zanzibar , sea farming is almost non existent in other African countries,

Japan, Republic of Korea, China, the Philippines and Taiwan (China) have highly successful culture of seaweeds . While Philippines produce *Eucheuma* seaweed mainly for industrial purposes, Japan is the largest producer of edible seaweeds. Major species are *Porphyra* and *Underia*. China's main species include *Gracilaria*, *Laminaria*, *Underia* and *Porphyra*, which are cultivated on floating rafts. However, Chile in South America has done exceptionally well in seaweed cultivation and processing.

It is important that countries intending to initiate or intensify mariculture should suitably establish linkage among each other and also with the following institutions.

Shandon College of Oceanology, China,

Korea Fishing Training Centre at Pusan, South Korea

Bureau of Fisheries of the National Fisheries Administration in Seoul, South Korea

School of Fisheries and Ocean Sciences, University of Alaska, Fairbanks, Alaska

The above suggestions of possible linkages between these institutions should serve as a model for establishing such linkages with other institutions in Africa and also with relevant institutions in South America. It is important that every country should take advantage of the technology and skill available in other country for the growth of fisheries sector in the developing countries.

9. References

Abidi, S.A.H. and N.K. Thakur (1998) Development of Professional Competency in our Fisheries Work Force- Changing Needs of 21 st Century, *Vision on Indian Fisheries of 21st Century*, CIFE, Mumbai

ADB/NACA (1998). Aquaculture Sustainability & the Environment. *Report on a Regional Study and Workshop RETA 5534* Asian Development Bank. Bangkok*: NACA;491pp*

Ahmed, M., M. M. Dey and M. J. Williams (1999). Assessment of the impact of Aquatic resources research by ICLARM: Scope and Methodologies. *Naga, the ICLARM Quarterly*, April/22(2) June :3-9

Andrew, M. (1991). Simple Methodology for Biological Assessment in River Ganga. *Proc. Work. Trg. Biomon. R. Ganga,* Thames Water International, UK. 1991, pp 47-49.

Asian Fisheries Society, (1988). Fisheries education and training in Asia. Workshop Proceedings. *Asian Fish Soc. Spec. Pub 1.2,* 162 p. Asian Fisheries Society, Manila, Philippines.

Balarine, J.D. (1988). System in use and species cultivated. *Village Level Aquaculture Development* in Africa Page 1-24. Commonwealth Secretariat, Malborough House, Pall Mall, London SW 145 HX.

Barrett, S.C.H. (1989). Water weed Invasion. *Scientific American* October 1989.

BFRI (1999) Five Year Research Master Plan (2000-2005), 26 August 1999, BFRI, Mymensingh.

Biradar, R.S. (2002) Course Manual on Fisheries Statistics. Second Edition CIFE, Mumbai. Pp 378.

Brown, L.R. and Wolf, E.C, (1984). Soil erosion: quiet crisis in the world economy. *Wold Watch paper* 60, 1984.

Bullock, A.M. and Cuts, R.P. (1985). The impact of the solar ultraviolet radiation upon the skin of rain bow trout, *Salmo gairdneri* Rishardson farmed at high altitude in Bolivia. *Journal of Fish Disease,* 1985, 8, 263-272.

Chua Thia-Eng. (1986). The need of review Fisheries Education Systems in Asia, *NAGA*, 9(4): 13-14.

CIFE (2000). Brainstorming session on "Manpower requirements and HRD in fisheries sector" held at CIFE, Mumbai, India 20-21 October 2000.

Collins, J. (1993) The Fisheries Branch Library of FAO, *FAO Aquaculture Newsletter*, April 1993 Number 3 PP 9-10

Collion, M.H. and A. Kissi (1995). Guide to program planning and priority setting. RMG No. 2e. International Service for National Agricultural Research, The Hague : ISNAR.

Craig, J.F. : Larger Dams and Freshwater Fish Biodiversity, prepared for Thematic Review 11.1: Dams , Ecosystem Functions and Environmental Restoration, World Commission of Dams.

Drinkwater, K.F.& K.T. Frank (1994) Effects of river regulation and diversion in marine fish and invertebrate. *Aquatic Conservation, Freshwater and Marine Ecosystem 4:135-151.*

Falconi, C.A. (1993). Economic Evaluation, *Monitoring and Evaluating Agricultural Research: A Source book*, PP 65-73. International Service for National Agricultural Research, The Hague, The Netherlands.

FAO/UNDP (1984) National Agriculture Research Report of An Evaluation Study in Selected Countries, FAO. Rome.

FAO, 2000. The Status of World Fisheries and Aquaculture 2000, Fisheries Department, FAO of the United Nations, Rome, Italy.

FAO/UNDP (1990) *Planning and managing forestry research: guideline for managers*, FAO Forestry Paper 96, FAO, Rome, Italy.

Gerard, V.A. (1984). Physiological effects of EI Nino on giant kelp in Southern California *Mar. Biol. Lett,* 5: 317-322.

Glynn, P.W. (1984). Widespread coral mortality and the 1982-83 EI Nino warming events. *Environ. Conserv.* 11 (2): 113-146.

Gopa Kumar, K (1998)Value addition and Product Development in Fisheries –Need for 21st Century in, *Vision on Indian Fisheries of 21st Century,* CIFE, Mumbai.

Grindle, Merilee S. (1997). Divergent Culture? When Public Organizations Perform Well in Developing Countries, *World Development, 25 (4): 481-495.*

Halweil, B. (2001). Environmental Intelligence. *World Watch*, 14(5): Sept/Oct 2001.

Hinrichsen, D.(2003) A Human Thrust. *World Watch*, Jan/Feb 2003

Horton D. , P. Ballantyne, W. Peterson, B. Uribe, D. Gapasin, K. Sheridan (1993). Monitoring and Evaluating Agricultural Research: A Source book, International Service for National Agricultural Research, The Hague, The Netherlands, PP 203.

Hubbs, C.L. (1984). Changes in the fish fauna of Western North America correlated with changes in ocean temperature. *J. Mar. Res.* 7: 459-482.

Hunter, J.R., Kaupp, S.E. & Taylor, J.H. (1981). Effects of solar and artificial UV-B radiation on larvae of northern anchovy *Engraulus mordax. Photo-chemistry and Photo-Biology* 29, 325-338.

ICRISAT (1992). ICRISAT Medium Plan 1994-98, Vol.I. International Crops Research Institute for the Semi-Arid Tropics, Patancheru, India.

ICLARM, (1986). Fisheries and aquaculture training opportunities on the ICLARM database as of November 1986, *NAGA, 9* (4) : 18-29

Janssen, W. and A. Kissi. (1997). Planning and priority setting for regional research. RMG Guidelines No. 4. ISNAR: The Hague.

Joshi, P.K. and M.C.S. Bantilan (1997) Institutionalization of Research Prioritization, Monitoring and Evaluation. *Workshop on Institutionalization of Research Prioritization, Monitoring and Evaluation*, 21-23 July 1997 at the National Center for Agricultural Research and Policy, New Delhi

Kapetsky, J. M. (1994). A strategic assessment of warm-water fish farming potential in Africa, *CIFA Technical Paper-27*, FAO, Rome.

Kaushik, S. J. (2002) Challenges and trends in fish nutrition. *The Sixth Indian Fisheries Forum*, 17-20 Dec., 2002. Pp 136-142. CIFE, Mumbai, India.

Kesteven, G.L.(1966) Organization and Administration of Fisheries Research. Working paper in the Meeting of Organization and Administration of Fisheries Research, *FAO Working Paper FAS/Bp/66/3*, FAO. Rome

Khoshoo. T.N (1998) Role of Science and Technology in Environmental Management. *The Shaping of Indian Science*. Vol. III : 1982-2003. Pp 1387-1566 University Press (India) Pannal Limited, 3-5-819 Hyderabad, Hyderabad-500 029

Kirby, A (2004) Hungry world 'must eat less meat' BBC. News on line 15 August 2004, U.K.

Linden, E. (2000), State of the Planet, Condition Critical, Times Earth Day, *Time*, April-May, 2000, New York, PP 18-24.

Lorenzen, K.(1995). Population dynamics and management of culture-based fisheries. *Fisheries Management and Ecology*, 2:61-73

Mazid, M.A. and V. R.P.Sinha (2000 a) Coastal Aquaculture Planning and Development in Bangladesh. *In Seminar on Rural and Coastal Aquaculture in Poverty Alleviation*, BFRI, March 2000:118-128p

Mazid, M.A. and V.R.P. Sinha (2000 b) Rural Aquaculture Planning and Development in Bangladesh. *In Seminar on Rural and Coastal Aquaculture in Poverty Alleviation*, BFRI, March 2000; 23-33 p

Mc Cully, P. (2001) Stream of Consciousness, *World Watch*. Jan/Feb 2001; 36-39

Mc Ardle, J and A.M.Bullock. (1987) Solar ultraviolet radiation as a causal factor of 'Summer Syndrome' in cage reared Atlantic salmon, *Salmo salar* L, A clinical and histopathological study. *Journal of Fish Disease*, 1987(10): 255-264.

Mruthyunjaya (2001). Impact of Farm Technologies lecture delivered in. *National Symposium on Fisheries Technologies and their Commercialization 11-12 Jan. Indian Fisheries Association CIFE, Versova*, Mumbai, India.

Nierenberg, D. (2001) Toxic Fertility. *World Watch*, Vol. 14(2): 30-38 March- April, 2001.

Nelson,J.S.(1994). *Fishes of the World*, 3rd Ed. John Wiley &Sons. Inc. New York, 620pp

Nair, N.B. (1998). Resource Management – A Perspective Plan in *Vision on Indian Fisheries of 21st Century*, CIFE, Mumbai

Norton, G.W. (1993). Scoring Methods, *Monitoring and Evaluating Agricultural Research: A Source book*, PP 102-168. International Service for National Agricultural Research, The Hague, The Netherlands.

Olah, J. and V.R.P. Sinha (1984). Principles and methods of monitoring of perennial undrainable pond ecosystem in tropical monsoon lands. *Aquacultura Hungarica,* 4: 103-110.

Olah, J. and V.R.P. Sinha (1986). Energy cost in carp farming systems. *Aquacultura Hungarica*, Vol. V. pp. 113-118.

Postal, S. (1991*) Oasis, World Watch Series,* Washington

Qasim, S.Z. (2002). Environmental Problems of Estuaries in India. *The Sixth Indian Fisheries Forum*, 17-20 Dec., 2002, CIFE, Mumbai, India.

Paroda, R. S. (1998) Food, Nutrition and Environmental Security. *The Shaping of Indian Science*. Vol. III : 1982-2003 pp 1999-2011 University Press (India) Pannal Limited, 3-5-819 Hyderabad, Hyderabad-500 029

Renner, Mitchael (2001) Going to work for Wind Power, *World. Watch* Jan/Feb.2001

Runyan, C. (2000). All the Wild Rivers, *World Watch*. Jan/Feb 2001; 31-33

Ryan J.C. (2001). Indonesian's Coral reef on the Line, *World Watch*, 14(3): 13-19. May-June 2001

Saxena, R, Suresh Pal and Mruthyunjaya (2002), A framework for setting Priority research. *The Sixth Indian Fisheries Forum*, 17-20 Dec., 2002, CIFE, Mumbai, India. Pp 82-90.

Shetty, H.P.C. (2000) Fisheries Education in the Asia Pacific Region, *The Sixth Indian Fisheries Forum,* 17-20 Dec., 2002 Pp 17-24, CIFE, Mumbai, India

Shuman,J.R. (1995). Environmental considerations for assessing dam removal alternatives for river restoration. *Regulated Rivers: Research and Management 11:249-261*

Sinha, V.R.P. (1980). Hydration of female spawners of carps during hypophysation. *Hydrobiologia, 72:* 193-196.

Sinha, V.R.P. (1981). Integrated synergic approach to aquaculture. *Resource Management and Optimization.* New York, USA 1 (4): 331-341.

Sinha, V.R.P. (1990). Paddy-fish production System. *Aquaculture Productivity* (ed.V.R.P. Sinha and H.C. Srivastava) Oxford and IBH Co. Pvt. Ltd. PP 313-381.

Sinha, V.R.P. (1990) Inland Fisheries Development – Planned management needed. *Hindu Survey of Indian Agriculture.*pp 199-201. 1990.

Sinha, V.R.P, (1991). *A Compendium of Aquaculture Technologies for Developing Countries.* Oxford & IBH Publishing Co. Ltd. Delhi, 115 pp

Sinha, V.R.P. (1993). Water resource and other modern developments and fisheries their interplay on environment *FAO Field Document 3. Short Term Training on Fisheries and Environment,* Mymensingh, Bangladesh.

Sinha,V.R.P. (1994). Introductory remarks on sustainable development of marine fisheries resources in Bangladesh*: Proc. of the Seminar on Sustainable Development of Marine Fisheries Resources in Bangladesh,* (Ed. By M.A. Mazid, V.R.P. Sinha and Md. Kamal). BFRI and FAO,

Sinha, V.R.P. (1996a) Country Report on the Status of Shrimp Culture, Trade and Environment in India, Mimeo 31 PP, FAO/INFOFISH, Kuala Lumpur, Malaysia

Sinha, V.R.P. (1996b) Country Report on the Status of Shrimp Culture, Trade and Environment in Bangladesh, Mimeo 26pp, FAO/INFOFISH, Kuala Lumpur, Malaysia

Sinha, V.R.P. (1997). Report on Potentialities, Policies and Packages of Training for Aquaculture Development in Africa, FAO/Regional Office for Africa, Accra, Ghana

Sinha, V.R.P. (1999). *Rural Aquaculture in India,* FAO/RAPA Publication no. 1999/21, Bangkok,

Sinha, V.R.P. (2000 a) Consultancy Report on Research Planning and Management, ARMP, BARC, Dhaka

Sinha, V.R.P. (2000b) Consultancy Report on Fish Reproductive Physiology, ARMP, BARC, Dhaka

Sinha, V.R.P. (2001). Consultancy Report on Research Planning and Management (Part II), ARMP, BARC, Dhaka.

Sinha, V.R.P. (2002). Aquaculture and Aquatic Environment Sustainability. *The Sixth Indian Fisheries Forum*, 17-20 Dec., 2002. Pp 25-32,CIFE, Mumbai, India

Sinha, V.R.P. KK. Ghosh & RS Biradar.(1991) Fisheries Education and Training in developing countries in 2000 A.D.(Mimeo 40 pp), Paper presented at the *World Fisheries Congress*, 14-19 April, 1991, Athens, Greece.

Sinha, V.R.P. Jhingran, V.G. and Ganapati, S.V. (1974). A review on the spawning of Indian major carps. *Arch. Hydrobiol.,* Stuttgart, Germany 73 (4): 518-536.

Sinha, V. R. P., and Jones, J. W. (1966). On the sex and distribution of the Freshwater eel *(Anguilla anguilla) J. Zool.,* London, U.K. 150: pp. 371-385.

Sinha, V.R.P. and E. Woynaorovich (1984). Spawning of cultivated finfish in Asia and Far East. *Agriculturl Hungarica,* Szarvas (Hungary), 4: 43-49

Singh Kohli, M.P. (1998) Manpower Requirements in Indian Fisheries Sector, *Fishing Chimes* ,18(2) :42-46

Sugunan, V.V. and M. Sinha (2000) Guidelines for Small Reservoir Fisheries Management, Bull. No 93, Central Inland capture Fisheries Research Institute, *(ICAR)*, Barrackpore-743101, West Bengal, India

Swaminathan, M.S.(1995) Agricultural Scientist's Perceptions on National Water Policy, National Academy of Agricultural Sciences, New Delhi

Swaminathan, M.S.(1996) Towards an Ever-green Revolution. Souvenir,2nd International Crop Science Congress, 17-24 November 1996, New Delhi, India pp-29-35,

Tacon, A (1992) The Food and Feeding of Farmed Fish and Shrimp *FAO Field Documents 1973-1991*, FAO, Rome

Ward, D.A. (2002) *Water Wars – Drought, Flood, Folly, and the Politics of Thirst,* Riverhead Books, New York ,294pp

World Bank (19991) Fisheries and aquaculture research capabilities and needs in Africa, *World Bank Technical Paper Number 149, Fisheries Series*, Washuington, DC.

World Watch (2002) Environmental Intelligence, *World Watch* March/ April 2002:8

World Watch (2002) Commentary, *World Watch* March/April 2002:11

Zingde, M.D.(1998) Health of the Coastal Marine Environment and Fisheries Development in India in *Vision on Indian Fisheries of 21st Century,* CIFE, Mumbai, India

Annexure I

Outline of Training Programs for different levels of Participants

A. Training Programs on certain important technologies for Farmers (Sinha 1997)

1. Carp seed production in hatchery and nursery raising

Objective

To train farmers and entrepreneurs to properly raise brood stock and undertake seed production.

Course content

The following topics should be explained to participants with suitable demonstrations.

- Selection of brood stock, raising, stocking density, feeding, breeding cycle, identification and selection of the mature mother stock.
- Secondary sexual character, spawning habit, breeding behavior, courtship and spawning.
- Induced breeding through hypophysation, extraction of pituitary gland, preparation of pituitary extract, intra muscular injection, hydration, gain in weight after injection of pituitary; use of chemicals as ovulating agents.
- Stripping of males, fertilization of eggs, water hardening, care of eggs, incubation chamber/glass jar/plastic pool.
- Prophylactic measures in the hatchery.
- Larval feeds, their preparation, feeding.
- Hatching.

- Nursery preparation, eradication of weed, insects and mollusks, fertilization and stocking, duration of raising, stocking density.
- Feeding.
- Harvesting, conditioning/acclimatization, oxygen packing.
- Method and mode of transportation.
- Economics of input and output.

2. Major carp, tilapia and catfish culture in ponds.

Objective

To train farmers and entrepreneurs to undertake improved fish culture practices.

Course Content

The following topics should be explained to participants with suitable demonstration.

- Principal of fish culture, selection of suitable species, food chain in the pond, role of sunlight and oxygen, role of vegetation in ponds, desirable macrophytes, common, submerged, floating and emergent weeds and their role in the pond ecosystem.
- Photosynthesis, production and consumption of oxygen in pond, how to increase oxygen in the water, symptoms of stress in fish because of low oxygen.
- Algal bloom and method of control, manual, biological and chemical.
- Application of lime in the pond, advantages of lime application, doses, litmus paper test.
- Composting of manure, amount and frequency of manure depending upon the nutrient status, available facilities if any for soil and water testing, advice on quality and quantity of inorganic fertilizers, recycling of pond bottom nutrient.
- Fingerling stocking, time of stocking, size of stocking, stocking density, advantage of polyculture, different proportion for stocking.
- Need for feeding the fish, effects of environmental factors like temperature, oxygen, over crowding, different types of artificial feeds; agricultural

wastes and by products like oil cakes and mill residue, sugarcane wastes, feed of animal origin like slaughter house refuge, fish meal, trash fish, method of feeding, broadcasting, platform feeding, feeding trays and feeding schedule.

- Maintenance of water and soil quality for fish health and public health, causes of malaria and filaria and their control measure by fish.
- Basic knowledge of sanitation and fish health, cause of fish health problem, over stocking, malnutrition, unfavorable soils and water condition. Fish kill by weather conditions and depletion of oxygen because of algal bloom or bacterial bloom. Common fish disease, prevention and cure, method of treatment, dose and frequency of treatment.
- Harvesting, different method of harvesting, time of harvesting, duration of harvesting, complete single harvesting, or multiple harvesting. Transportation and marketing.
- Economics and profitability of fish culture compared to agriculture and livestock farming.

3. Rice-Fish Culture

Objective

To train rice farmers in rice fish culture integrated farming system.

Course Content

The following topics should be explained to the participants with suitable demonstrations.

- Principle of rice-fish culture. Rice fields as a fish pasture ground. Role of fish in supply of oxygen and nutrient to paddy. Common fish fauna in rice fields, desirable species, their food habit. Common insects in the field, their role and control.
- Preparation of rice field for raising of the seed or undertaking fish culture. Strengthening of dykes/bunds, determination of width and height of bunds, level of water required, construction of trenches, enclosures and fish hide out, water control structure, manipulation of water level during and after weeding.

- Method of safe application of pesticides and fertilizers. Measures to reduce the effect of pesticides and fertilizers.
- Measures for protection from predators.
- Stocking of fry for fingerling production, stocking density, and period of raising.
- Harvesting, packing and method of transport. Profitability of seed raising.
- Stocking for fish culture, suitable species, their stocking proportion, stocking density, duration of culture.
- Method of harvesting. Fabrication of suitable traps and gears. Profitability of rice-fish culture.

4. Integrated Fish-Livestock Farming

Objective

To train livestock farmers, entrepreneurs and aquaculturists to undertake integrated fish-livestock farming.

Course Content

The following topics should be explained to the participants with suitable demonstrations.

- Principles of integrated fish-livestock farming. Different methods of integration. Relationship between different production systems.
- Designs and construction of integrated fish farm. Selection of plants which should be cultivated on the farm and pond embankment, their farming techniques.
- Construction of the bird house.
- Manurial value of pond mud, and bird dropping. Plant and animal waste treatment, method of decomposition and composting.
- Preparation of the pond, selection of the species, stocking density and proportion of different species, their culture techniques.
- Depending on the nutrient status of soil and water, the amount of organic manure required for fish culture and accordingly calculation of the number of birds to be raised on the embankment.

- Duck/chicken rearing, egg production and marketing. Vegetable/crop production and marketing. Fish harvesting and marketing. Profitability of integrated fish farming.

5. Cage Culture

Objective

To train aquaculturists and entrepreneurs in cage culture technology.

Course Content

The following topics should be explained to the participants with suitable demonstrations.

- Principles and advantages of cage culture.
- Different types of cage, floating cage, rigid and flexible cage self supported and raft supported. Construction of cage, framework, support, selection of materials for framework. Mooring and anchoring, selection of materials for mesh netting.
- Selection of site and installation of cage, construction of water breaking structures.
- Selection of fish species, procurement of seed, stocking rates.
- Different types of feed, feeding of fish, feeding frequency and methods. Maintenance and clearing of nets. Prevention and control of diseases and predators.
- Selection of size at the time of harvest, harvesting. Profitability of cage culture.

6. Pen culture

Objective

To train aquaculturists and entrepreneurs s to undertake pen culture.

Course Content

The following topics should be explained to the participants with suitable demonstrations .

- Principles and advantages of Pen culture. Different types of pen, rigid pen/flexible pen.

- Design and construction, shapes and dimensions of pens, nursery and fattening pens.
- Selection of species of fishes either for monoculture or polyculture.
- Procurement of seeds, preparation of nursery, stocking rates, fertilization and feeding of fish, types of feed, method and frequency of feed.
- Predators and disease control.
- Monitoring of growth of fish and selection of optimum size of fish for harvesting.
- Transport and Marketing of fish. Profitability of pen culture.

7. Shrimp Culture

Objectives

To train aquaculturists and entrepreneurs in shrimp farming.

Course Content

The following topics should be explained to the participants with suitable demonstrations.

- Principle of shrimp farming. Types of shrimp farming, traditional, extensive, semi-intensive and intensive.
- Site selection, availability of good quality fresh and brackishwater, soil quality, land elevation, salinity, temperature, tidal amplitude. Design, construction and maintenance of farm pond.
- Pond preparation, cleaning of pond, eradication of predators, sun drying, ploughing and leveling of pond bottom. Liming for soil pH correction. Application of organic and inorganic fertilizers.
- Stocking, size/age of PL / density.
- Pond management, aeration and change of water. Stock management and manipulation. Production and maintenance of natural food, plankton production.
- Feeds and feeding, different types of feed. feeding rate and frequency.
- Water quality management, plankton bloom, water temperature, pH, control of disease and surveillance on shrimp health.

- Harvesting and post harvest handling of shrimps. Profitability of shrimp farming.

B. Outline of Certain Important Areas of Training for Extension Officers at the Secondary Level

1. Carp, tilapia and catfish Culture and Seed Production

Objectives

To train extension workers in carp, tilapia and catfish culture and seed production.

Major aspects to the covered during Training Course

- Principle of fish culture.
- Selection of brood stock and their management.
- Breeding and seed production.
- Nursery management.
- Harvesting, Transport and Marketing.
- Economics of seed production of carp, tilapia and catfish.
- Economics of carp, tilapia and catfish culture.

2. Rice-fish Culture

Objective

To train extension workers in rice-fish culture technology.

Major aspects to the covered during Training Course

- Principle of fish-rice production system.
- Water management in low-lying and upland paddy fields.
- Paddy field ecology.
- Fish seed raising technology in paddy field.
- Fish culture technology in paddy field.
- Inter-relationship between paddy and fish.

- Harvesting techniques.
- Economic evaluation.

3. Integrated Fish Farming

Objective

To train extension workers in Integrated Fish Farming Technology.

Major aspects to the covered during Training Course

- Principles of integrated fish farming.
- Inter-relationship between different sub systems of production.
- Integrated management of fish-crop farming.
- Integrated management of fish-livestock-crop production.
- Models of fish-livestock-crop integration.
- Economic evaluation of integrated fish farming.

4. Shrimp and Prawn Hatchery Technology and Farming

Objective

To train extension workers in Shrimp and Prawn Seed Production and Shrimp Farming.

Major aspects to the covered during Training Course

- Principle of shrimp and prawn seed production and farming.
- Prospecting of mother stock of shrimp and prawn.
- Hormone manipulation and induced maturity in shrimp.
- Controlled breeding of shrimp and prawn.
- Maintenance and culture of live food organism.
- PL raising.
- Different type of hatchery maintenance.
- Hatchery management for shrimp/prawn.
- Shrimp culture technology.
- Prawn culture technology.

- Farm Management. Harvest. Post harvest. Product Development and Marketing.

5. Cage/Pen Culture

Objective

To train extension workers in cage/pen culture technology.

Major aspects to be covered during Training Course

- Principle of cage/pen culture
- Suitable species for cage/pen culture
- Fabrication, maintenance and installation of cages
- Fish husbandry in cages/pens
- Economics of cage/pen culture of different species.

6. Hilsa Fisheries Management and Conservation

To train extension officer on different aspects of Hilsa fisheries management and conservation.

- Biology & ecology of hilsa
- Population dynamics and recruitment pattern of hilsa
- Hilsa fisheries management
- Conservation & rehabilitation of Jatka
- Hilsa fisheries development

C. Brief Outline of Some of the Important Training Areas for Tertiary Level Participants (Researchers, University Teachers and Social Scientists dealing with Aquaculture).

1 Pond dynamics and maintenance of proper soil and water quality.

The following aspects along with suitable experimentation need to be covered under the training program.

1. Primary productivity utilization and judicious use of fertilizers and feeds.
2. Understanding role of microbes in pond ecosystem.

3. Recycling of nutrients of the pond.
4. Role of chemicals and micronutrients in soil and water quality improvement and productivity enhancement.
5. Proper water quality, its relationship with health and hygiene of fish and human being.

2. Controlled breeding and seed production of carp, catfish and tilapia

The following aspects along with suitable experimentation need to be covered under the training program.

1. Principle and procedure of management and maintenance of the broodstock of carp, catfish and tilapia.
2. Understanding the eco-physiological and hormonal aspects of the breeding of fish to have better control on reproduction.
3. Understanding of the spawning requirement and the principle and procedure of induced breeding of fish.
4. Principle and procedure of hatchery management, production of healthy seeds, their transportation and the economics of fish seed trade.

3. Fish Health Management

The following aspects along with suitable experimentation need to be covered under the training program.

1. Understanding diagnostic procedure for viral, bacterial, mycoses and algal disease and protozoan and metazoan parasitic diseases.
2. Understanding the mechanism of environmental stress and environment induced diseases,
3. Methods of fish health monitoring system; immunoprophylatic and other preventive measures and methods of control of fish diseases.

4. Fish Feed and Nutritional Studies

The following aspects along with suitable experimentation need to be covered under the training program.

1. To adopt standardized nutritional research methodology and methods of assays.

2. To understand the nutritional requirement of different life stages of different fish and also of nutritional value of natural feeds.
3. To compile and collate the information on the availability of locally available by product of agro-industry for formulation of the balanced and cost effective feed for the brood stock and of seed.

5. Fish Stock Improvement and Genetical Research

The following aspects along with suitable experimentation need to be covered under the training program.

1. Documentation and evaluation of fish genetic resource for genetical up gradation.
2. Principles and procedure of a wide range of manipulatory techniques in genetical studies and adoption of the standardized genetical research methodology and methods of assay.
3. Development of genetically superior strain with better cultivable qualities by genetic manipulation to maximize the aquaculture output.

6. Marketing of Aquaculture Products

The following aspects along with relevant case studies should be covered under the training program.

1. Selection and assumption of parameters to estimate the existing and future market potential of the species in different product forms and in various markets.
2. Evaluation of the marketing constraints and the means to circumvent those.
3. Methodology to identify the correct marketing strategies for different species.
4. Marketing and advertisement for promotion of new aquaculture products.

7. Socio-economics of Aquaculture Development

The following aspects along with relevant case studies should be covered under the training program.

1. Methods to assess the socio-economic impacts of aquaculture on food security, employment, income, and foreign exchange earning.

2 Evaluation of the economics of aquaculture operations in different scales and types of management.

3 To identify the social and economic constraints for aquaculture development.

8. Inland Fish Stock Assessment

The following aspects along with suitable experimentation, documentation and field visits should be covered under the training program.

1. To understand the relevance of collecting and collating the relevant data on inland fish and fisheries resources.
2. To understand the changes taking place in the inland fisheries because of human intervention and modern development.
3. To understand how to undertake studies on population dynamics and stock assessment of fish through different methods and models.
4. Principle and parameters to evolve catch monitoring system of inland fishes.
5. Relevance of Remote Sensing and GIS in Inland Fish Stock assessment studies.

9. Open water Fisheries Management

The following aspects along with suitable case studies should be covered under the training program.

1. To understand how to undertake studies on limnological and ecological aspects to assess and monitor productivity.
2. To understand different aspects of open water fisheries management.
3. To conduct studies on important species on their distribution, migration and environmental preference, their modes of reproduction and reproductive seasons; their food types and feeding inter-relationship and major predators.
4. To understand how to undertake studies on population dynamics and stock assessment of fish through different methods and models.
5. Relevance of Remote Sensing and GIS and GPS in open water fisheries management.

10. Culture based Inland Fisheries Management

The following aspects along with suitable case studies should be covered under the training program

1. To understand the basic principles of culture based fisheries management
2. To understand the package of practices in terms of suitable species, size, stocking density and other management measures for culture-based fisheries.
3. To understand the different aspects of rehabilitations of inland fisheries.

11. Aquatic Pollution and Fish Toxicology

The following aspects along with suitable experimentation, field visits and case studies need to be covered under this training program.

1. To understand the method of monitoring the aquatic pollution and its impact on fisheries.
2. To understand how to undertake studies on fish toxicology with different pesticides and monitor status of new pesticides and their effects on the different stages of life cycle of fish.

12. Biodiversity, Fish Genetics Resources and Conservation

The following aspects need to be covered along with field visits under the training program.

1. To understand the methodology of assessment of the aquatic biodiversity and to identify threatened fish species and fish food organisms.
2. To undertake collection, identification and evaluation of fish genetic resources.
3. Establishment and management of fish sanctuary.

13. Fish behavior and Fishing Technology

The following aspects need to be covered along with suitable survey and field visits under the training program.

1. To understand the use and misuse of different gear and crafts used in different types of fisheries.

2. To select and motivate fishermen to use desirable craft and gears.
3. To understand basic fish behavior with different types of craft and gears.

14. Fish Reproductive Physiology

The following aspects need to be covered along with suitable experimentation and hatchery visit.

1. Importance of studies on reproductive physiology, ecophysiology and endocrinology in breeding of fish and shrimp for culture and quality seed production.
2. Methods of characterization of fish gonadotropins, endocrinological aspects of breeding, pheromones and fish reproduction, and eco-physiology of fish reproduction particularly in terms of environmental condition, water quality and spawning of commercially important fishes and shrimps.

15. Brackish water Aquaculture

The following aspects need to be covered along with suitable experimentation and farm visit.

1. Principles and procedure of brackishwater fish and shrimp culture and the recent advances in the techniques for the commercial culture of potential brackish water species of fish and shrimp.
2. To select parameters and conduct environmental impact studies on coastal aquaculture development.

16. Estuarine Ecology

The following aspects need to be covered along with suitable field visit.

1. To select parameters and undertake environmental impact studies of modern development on estuarine ecology.
2. To understand the methodology to study the biological diversity of estuarine, mangrove and coastal ecology and identify the threatened species of fish and prawn.

17. Aquaculture Engineering

The following aspects need to be covered along with suitable field visit.

1. To study the engineering problems of freshwater and coastal aquaculture and the methods of circumventing them.
2. To know how to establish cost effective hatchery for fish and shrimps.
3. To understand the principles of farm design, water structure and other associated structure for efficient farm management.

18. Marine Fish Behavior and Fishing Technology

The following aspects need to be covered along with suitable field visit.

1. To understand the behavior of important marine species, particularly their behavior in response to fishing gear.
2. To know about marine fishing technologies, including fishing craft, fishing gear and on-broad facilities, and in particular development of improved fishing technology for the marine artisanal fisheries.

19. Mariculture

The following aspects need to be covered along with suitable field visit.

1. Principles of mariculture and the techniques for the commercial culture of potential mariculture species.
2. To know the recent advances on reproductive physiology, genetics, nutrition and environmental requirements of commercially important marine fin fish and shell fish.
3. To know the recent advances in seaweeds cultivation and in their processing,

20. Processing, Product Development and Quality Control

The following aspects need to be covered along with suitable field and processing plants, visit.

1. To know the nutritional qualities, physical, microbiological and biochemical properties of important fish and understand the methodology of their post-harvest handling, preservation and processing of important marine species.

2. To understand the different methods of value-addition of marine foodstuff and know how to have the quality control of fish and fish products.

3. To know the Hazard Analysis and Critical Control Point and undertake relevant experimentation.

21. Marine Pollution and Fisheries

The following aspects need to be covered along with suitable field and on board visit.

1. To understand physical and chemical oceanography with particular attention to the water movements and characteristics, relevant to fish population.

2. To know the biological oceanography and its production systems that support marine fisheries resources.

3. To understand the method of monitoring marine pollution and its impact on marine fisheries.

22. Marine Fish Stock Assessment

The following aspects need to be covered along with suitable field, landing centers and on board visits.

1. To understand how to assess resource biomass, catch and fishing efforts and to gather other fisheries statistics on marine fisheries.

2. To know how to monitor the population structure and dynamics of marine resource population, including studies of their age and growth rate, distribution, migrations and environmental performances; their modes of reproduction and reproductive seasons; their food types and major predators.

3. To understand how to undertake studies on population dynamics and stock assessment of fish through different methods and models

4. Relevance of Remote Sensing and GIS in marine fish stock assessment.

23. Marine Fisheries Economics

The following aspects need to be covered along with suitable case studies and field visit.

1. To understand the economic and socio-economic aspects of marine fisheries and to know how to allocate the limited marine fisheries resources between small-scale fishing communities and industrial fisheries to avoid the conflict between them.
2. To select the parameters for socio-economic impact studies of marine fisheries related policies and projects and undertake assessment of impact through some case studies.